AF001117

Sustainable Development and Quality Assurance in Higher Education

Palgrave Studies in Global Higher Education

Series Editors:

Roger King, School of Management, University of Bath, UK; **Jenny Lee**, Centre for the Study of Higher Education, University of Arizona, USA; **Simon Marginson**, Institute of Education, University of London, UK; **Rajani Naidoo**, School of Management, University of Bath, UK

This series aims to explore the globalization of higher education and the impact this has had on education systems around the world including East Asia, Africa, the Middle East, Europe and the US. Analysing HE systems and policy, this series will provide a comprehensive overview of how HE within different nations and/or regions is responding to the new age of universal mass higher education.

Titles include:

Michael Dobbins and Christoph Knill
HIGHER EDUCATION GOVERNANCE AND POLICY CHANGE IN WESTERN EUROPE
International Challenges to Historical Institutions

Zinaida Fadeeva, Laima Galkute, Clemens Mader and Geoff Scott (*editors*)
SUSTAINABLE DEVELOPMENT AND QUALITY ASSURANCE IN HIGHER EDUCATION
Transformation of Learning and Society

Forthcoming:

Christof Van Mol
INTRA-EUROPEAN STUDENT MOBILITY IN INTERNATIONAL HIGHER EDUCATION CIRCUITS
Europe on the Move

Lý Trần, Simon Marginson, Hoàng Đỗ, Quyên Đỗ, Trúc Lê, Nhài Nguyễn, Thảo Vũ, Thạch Phạm and Hương Nguyễn
HIGHER EDUCATION IN VIETNAM
Flexibility, Mobility and Practicality in the Global Knowledge Economy

---

Palgrave Studies in Global Higher Education
Series Standing Order ISBN 9781137348142 Hardback
(*outside North America only*)

You can receive future titles in this series as they are published by placing a standing order. Please contact your bookseller or, in case of difficulty, write to us at the address below with your name and address, the title of the series and the ISBN quoted above.

Customer Services Department, Macmillan Distribution Ltd, Houndmills, Basingstoke, Hampshire RG21 6XS, England

Sustainable Development and Quality Assurance in Higher Education

Transformation of Learning and Society

Edited by

Zinaida Fadeeva
United Nations University

Laima Galkute
Vilnius University, Lithuania

Clemens Mader
Leuphana University of Lüneburg, Germany
University of Zurich, Switzerland

and

Geoff Scott
University of Western Sydney, Australia

Selection and editorial matter © Zinaida Fadeeva, Laima Galkute, Clemens Mader and Geoff Scott 2014
Individual chapters © Respective authors 2014
Foreword © Kazuhiko Takemoto 2014
Foreword © Dzulkifli Abdul Razak 2014
Foreword © Daniella Tilbury 2014

All rights reserved. No reproduction, copy or transmission of this publication may be made without written permission.

No portion of this publication may be reproduced, copied or transmitted save with written permission or in accordance with the provisions of the Copyright, Designs and Patents Act 1988, or under the terms of any licence permitting limited copying issued by the Copyright Licensing Agency, Saffron House, 6–10 Kirby Street, London EC1N 8TS.

Any person who does any unauthorized act in relation to this publication may be liable to criminal prosecution and civil claims for damages.

The authors have asserted their rights to be identified as the authors of this work in accordance with the Copyright, Designs and Patents Act 1988.

First published 2014 by
PALGRAVE MACMILLAN

Palgrave Macmillan in the UK is an imprint of Macmillan Publishers Limited, registered in England, company number 785998, of Houndmills, Basingstoke, Hampshire RG21 6XS.

Palgrave Macmillan in the US is a division of St Martin's Press LLC, 175 Fifth Avenue, New York, NY 10010.

Palgrave Macmillan is the global academic imprint of the above companies and has companies and representatives throughout the world.

Palgrave® and Macmillan® are registered trademarks in the United States, the United Kingdom, Europe and other countries.

ISBN: 978–1–137–45913–8

This book is printed on paper suitable for recycling and made from fully managed and sustained forest sources. Logging, pulping and manufacturing processes are expected to conform to the environmental regulations of the country of origin.

A catalogue record for this book is available from the British Library.

Library of Congress Cataloging-in-Publication Data

Sustainable development and quality assurance in higher education : transformation of learning and society / Zinaida Fadeeva, United Nations University, Japan; Laima Galkute, Vilnius University, Lithuania; Clemens Mader, University of Lüneburg, Germany; Geoff Scott, University of Western Sydney, Australia.
    pages cm.
    ISBN 978–1–137–45913–8 (hardback)
    1. Sustainable development – Study and teaching (Higher) 2. Education, Higher – Administration. 3. Education, Higher – Social aspects. 4. Quality assurance I. Fadeeva, Zinaida, editor of compilation.

HC79.E5S866825 2014
338.9′27—dc23   2014028171

UNU-IAS
Institute for the Advanced Study of Sustainability

Contents

List of Boxes	x
List of Figures	xi
List of Tables	xiii
Foreword Kazuhiko Takemoto	xiv
Foreword Dzulkifli Abdul Razak	xv
Foreword Daniella Tilbury	xviii
Acknowledgements	xix
Notes on Contributors	xxi
List of Abbreviations and Acronyms	xxx
Structure of the Book	xxxiii

1 Assessment for Transformation – Higher Education Thrives in Redefining Quality Systems   1
Zinaida Fadeeva, Laima Galkute, Clemens Mader and Geoff Scott

Part I Transformation of Higher Education in Changing Society: Implications for Quality Management

2 Rankings and the Reconstruction of Knowledge during the Age of Austerity   25
Ellen Hazelkorn

3 Linking Quality Assurance and ESD: Towards a Participative Quality Culture of Sustainable Development in Higher Education   49
Oliver Vettori and Christian Rammel

4 The Role of Assessment and Quality Management in
  Transformations towards Sustainable Development:
  The Nexus between Higher Education, Society and Policy   66
  Clemens Mader

Part II The Meaning and the Role of the Internal Quality Assurance and Its Interplay with External Quality Approaches in Supporting HE Sustainability Transformation

5 Drivers for Change in the Austrian University Sector:
  Implications for Quality Management   87
  Nadine Shovakar and Andrea Bernhard

6 A Quality Assurance System Based on the Sustainable
  Development Paradigm: The Lithuanian Perspective   114
  Laima Galkute

7 Quality System Development at the University of
  Graz: Lessons Learned from the Case of RCE Graz-Styria   131
  *Friedrich M. Zimmermann, Andreas Raggautz,
  Kathrin Maier, Thomas Drage, Marlene Mader, Mario Diethart
  and Jonas Meyer*

8 STARS as a Multi-Purpose Tool for Advancing
  Campus Sustainability in US   153
  Monika Urbanski and Paul Rowland

Part III Quality Management and Facilitating Sustainability Competences and Capabilities

9 Sustainability and Values Assessment in Higher Education   185
  Arthur Lyon Dahl

10 Educating Sustainability Change Agents by Design:
   Appraisals of the Transformative Role of
   Higher Education   196
   *Katja Brundiers, Emma Savage, Steven Mannell,
   Daniel J. Lang and Arnim Wiek*

11 Quality Management of Education for
   Sustainability in Higher Education   230
   Geoff Scott

12 Implementing Education for Sustainable Development in Higher Education: Case Study of Albukhary International University, Malaysia   255
   Salfarina Abdul Gapor, Abd Malik Abd Aziz, Dzulkifli Abdul Razak and Zainal Abidin Sanusi

Index   283

List of Boxes

0.1	Transformation, sustainability and higher education	xxxiii
1.1	Definitions	2
1.2	Science for sustainable development	7
1.3	ESD competences of educators	10
2.1	Major global rankings, 2014	29
2.2	Changes in our understanding of knowledge production	34
6.1	Examples of university mission statements	117
6.2	Quality criteria	122
8.1	About AASHE and STARS	154
8.2	Five methods for transforming campus sustainability through STARS	159
8.3	Sustainability in curriculum, research and student engagement	176
8.4	Sustainability in Planning, Administration & Engagement (PAE)	178
11.1	Reference points that can be used to validate learning outcomes in higher education learning programmes	239
11.2	Successful graduate studies	242

List of Figures

4.1	Iterative process of transformation enabled through quality management and assessment	76
5.1	Internal and external quality management tools as drivers for change	107
6.1	Preferences in defining the role of higher education by university teachers and employers	119
6.2	Interconnection between strategic management of the HEI and quality assurance	122
7.1	Quality management cycle of the University of Graz	137
7.2	System elements of the quality management cycle at the University of Graz	139
7.3	Transdisciplinary research at the RCE Graz-Styria	145
8.1	2014 analysis of average scores for all institutions under STARS Version 1	161
8.2	Analysis of STARS Version 1 data pertaining to maintained building space (*OP 1: Building Operations & Maintenance*) and new construction (*OP 2: New Building Design & Construction*)	172
8.3	Total emissions for all rated institutions under STARS Version 1.0 through 1.2, displayed in thousands metric tons carbon dioxide equivalent (CO_2e) in 2013	173
8.4	2013 analysis of change in energy consumption (in millions MMBtu) for all STARS Version 1 reports	174
8.5	2013 analysis of change in building space (in millions gross square feet) for all STARS Version 1 reports	174
8.6	2013 analysis of PAE subcategory scores for all institutions submitting under STARS Version 1	177
8.7	2013 analysis of average scores by category based on institution type for all institutions submitting under STARS Version 1	179

10.1	Illustration of how the five competences in sustainability relate to components of a sustainability research and problem-solving	200
10.2	Five phases for problem- and project-based, solution-oriented sustainability research courses and project-participant constellation per phase, with phase-lead in grey	202
10.3	Organization for team-based courses to support collaboration with project-clients and peer-learning	208
11.1	UWS framework for assuring academic standards and quality in L&T	236
11.2	Professional capability framework	241
12.1	Sustainable livelihood approaches	263
12.2	AIU/SLA framework for sustainability assessment	264
12.3	Sustainability Rating Assessment System (SRAS) for AIU/SLA programme	265
12.4	Sustainability rating for students' performance in the Kensiu project, Year 1 with targets for Years 2 and 3	278

List of Tables

2.1	Knowledge production accountability	35
2.2	Indicative mapping of institutional actions against selective rankings (2014)	38
6.1	Changing rhetoric on the role of higher education as reflected in the EHEA ministerial meetings	116
8.1	STARS ratings summary	161
8.2	Five highest- and lowest-scoring STARS subcategories	162
8.3	Five highest- and lowest-scoring STARS credits	163
8.4	Impact of environmental issues on college admissions	164
10.1	Analytical-evaluative framework adapted from Brundiers and Wiek (2013)	203
10.2	Preliminary learning framework for the ESS Major programme developmental evaluation	218
10.3	Current evaluation practices of ESS Major programme focusing on factual data	219
10.4	Proposed evaluation components of the new ESS enquiry framework based on qualitative aspects	220
12.1	Strengths and weaknesses of AUA, STARS and SEI	260
12.2	Teaching indicators for sustainability assessment	268
12.3	Research indicators for sustainability assessment	270
12.4	Community engagement for sustainability assessment	273
12.5	SRAS result for Kensiu Project, Year 1 (%) in relation to ESD-SD	277
12.6	Sustainability rating for Kensiu Project, Year 1 (scale)	277

Foreword

Quality of education, including that of higher education, has emerged as a key topic in the discussions on education for sustainable development (ESD). The publication of *Sustainable Development and Quality Assurance in Higher Education: Transformation of Learning and Society* is a contribution to the ongoing dialogue on integrating principles of sustainability to the educational systems. It is this system approach – at the level of higher education institutions and the policies and practices that govern them – that makes the book and practices that it highlights a unique contribution to the UN Decade of Education for Sustainable Development (DESD).

The analytical and development work that centres on quality management as a strategy for the transformation of higher education has been the focus of UNU-IAS attention for a number of years. It began with the action research project of ProSPER.Net, an alliance of leading universities in the Asia-Pacific region, where the member universities sought to develop an alternative university appraisal system that recognizes progress of universities in sustainability. While the strong emphasis on assessment as learning present in that earlier process remained our focus, the overall approach has become broader. Today, we use lessons from earlier experiences focused on sustainability to transform 'mainstream' educational systems. I hope that this publication, with its focus on the evolving and changing notion of higher education quality, organizational and policy interplays for the development of quality systems and the role of quality systems in facilitating competences, will add to this ambition of upscaling and mainstreaming sustainability into educational practices.

Dr Kazuhiko Takemoto
Director, United Nations University Institute for
the Advanced Studies of Sustainability (UNU-IAS)

Foreword

Albert Einstein has been quoted as saying, 'Not everything that can be counted counts, and not everything that counts can be counted.' In this context, education, especially education for sustainable development (ESD), seems to be straddled between the two dimensions.

Currently, the so-called key performance indicators' (KPIs) way of ascertaining outcomes by using quantitative measurement of performance in the education sector has created a 'bias' in the understanding of the quality concept – limiting it to 'what can be counted or measured' while marginalizing those outcomes that 'cannot be counted'. Without doubt the former is much easier to do (and as such is often preferred) compared to the latter. With this predicament at hand, the relevance and importance of this book, aptly entitled *Sustainable Development and Quality Assurance in Higher Education: Transformation of Learning and Society*, cannot be underestimated.

In other words, to understand 'quality' education by merely 'counting beans' is no longer acceptable, not only because of the growing challenges faced by education today – including the various external factors as described in the book's chapters – but more so given the multidimensional aspects of sustainability that need to be factored in at the same time. That is to say, without redefining the quality systems, ESD will continue to be left out of the mainstream as the cornerstone of education in achieving a sustainable future. A clear case in point is the question of the ethics and values that are vital in advocating ESD and creating new mindsets that respect and preserve the fragile eco-system in nurturing a lasting planetary co-existence.

Simply put, as the role of education changes towards a focus on more planetary, if not cosmic, dimensions, the focus on quality systems too must evolve to reflect and articulate this sea of changes. Closely related to this is the rooting of education within a specific cultural context and the worldview that blends with it. For instance, in the case of indigenous knowledge, its worldview is often more sophisticated in safeguarding the environment as part of a sustainable, if simple, lifestyle. This is well recognized by UNESCO (2000) which notes: 'Sophisticated knowledge of the natural world is not confined to science. Human societies all across the globe have developed rich sets of experiences and explanations relating to the environments they live in.' Lest the aspects of these 'other knowledges' are fully accounted for, the definition of 'quality' will continue to remain provincial, dominated in essence by Western/Northern worldviews which are less tolerant to other worldviews, especially of the colonized, if at all. This is clearly in contradiction to the basics of sustainability where the main thrusts are founded on the elements of collaboration, co-creation and embracing diversity in an inclusive way.

It is especially critical since indigenous knowledge and wisdom are still very vibrant in many 'developing' countries; it is imperative that this aspect be considered in earnest. That means the 'quality' dimensions must also be sensitive to these communities in promulgating the post-2015 development agenda. And, in doing so, it widens the acceptance of other human experiences. That is consistent with an understanding of 'quality' which is not dehumanizing nor does it obliterate or ignore values that are seemingly in conflict with those stereotyped as modern and/or contemporary as currently understood and practised.

The International Association of Universities (IAU) is fully supportive of the effort to develop a discourse around the themes that are covered by the book. For example IAU firmly believes and endorses the whole institution approach in developing and advocating ESD, underscored by a more holistic 'quality' system. It is through such a system that the approach to transformation must be realized by inter-linking the different entities at the various levels – local, national, regional and global on one hand, and people, planet, prosperity and politics on the other. Such a transdisciplinary mix will help create new and emerging vistas in shaping more adept higher education institutions in the Age of Sustainability. They are no longer the ivory towers of the past, but engaged institutions such as 'the humaniversity' described in Chapter 12. In the days of the Industrial Age universities were modelled to resemble the 'factory' metaphor (in fact, arguably the notion of 'quality' today is very much inspired by the same metaphor, so too the notion of KPIs); now they must be crafted to meet the demands of sustainability.

In light of this, the United Nations University's initiatives in establishing the Regional Centres of Expertise (RCEs) on Education for Sustainable Development launched, at the start of the UN Decade of Education for Sustainable Development in 2005 can be considered as a forerunner to sustainability-led university. Beginning with only seven pioneering RCEs, the numbers have grown, by the end of the decade, to more than 150, putting ESD on a global trajectory. The RCEs (several of which are active members of IAU) are without doubt well placed as test-beds for redefining and implementing a more encompassing quality system.

In its latest International Conference held in November 2014 at Iquitos, Peru, IAU renewed its commitment to Higher Education for Sustainable Development and agreed to develop new actions and strengthen current initiatives to respond proactively to the resolutions and recommendations put forth in the Iquitos Statement adopted at the end of the Conference. This includes, notably, the *whole institution approach* in translating sustainable development into institutional agendas, apart from building synergies and promoting collaboration in the search for effective and innovative solutions to solving today's as well as future sustainable development challenges. In addition, IAU commits to offering an open, interactive and collaborative forum for discussion and action, to raise awareness and advocate for change, while showcasing higher education institutions' activities from around the world and offering networking opportunities.

All this, however, is very much dependent on how 'quality' is perceived, (re)defined and eventually put to action to impact both tangible and intangible aspects inter-generationally. This will make ESD more enduring since it could be readily internalized not just for now but extending well into the future in line with the IAU aspiration to build a worldwide higher education community.

I believe this book has many invaluable insights, ideas, arguments and best practices with numerous examples and experiences that can stimulate and enrich many others to think afresh about the most relevant and suitable quality system to appropriately advance sustainable development in the higher education sector.

I would like to congratulate the editors and contributors in making this exciting volume possible: one that comprehensively delves into a long overdue subject area of immense significance to ESD as the UN Decade draws to a close.

Professor Dzulkifli Abdul Razak
President of the International
Association of Universities (IAU)

Foreword

Sustainability issues continue to rise up the global agenda, suggesting the need to rethink our present patterns of life. Our education and learning systems are critical to progress on sustainability, and higher education has produced ground-breaking research and excellent corporate practice in this area. However, a 2014 UNESCO report of higher education confirms that progress on the curriculum and whole-of-institution approaches to sustainability in higher education has been slow and piecemeal across the globe.

The imperative seems clear but, as the UN Decade of Education for Sustainable Development has revealed, the process of embedding sustainability in the curriculum raises significant challenges for academic leaders and teaching colleagues. This is not surprising, given that education for sustainable development reframes our understanding of quality and thus of the structures, frameworks and criteria which underpin its assessment. The experiences of colleagues from 35 countries engaged in the COPERNICUS Alliance-led project University Educators for Sustainable Development (UE4SD) testify to the scale of the challenge ahead but also the relationship between professional development and quality.

There is much work to be done to bring sustainability firmly into academic development and learning processes in higher education. This is where this publication, *Sustainable Development and Quality Assurance in Higher Education: Transformation of Learning and Society*, can make an important contribution. The chapters review ideas and initiatives in this area and raise questions about the quality of academic provision as seen through the ESD lens. It can promote dialogue, deepen understanding and build capacity for change. This book can help colleagues wrestle their way through the deeply entrenched unsustainable views of education, highlighted by the Rio+20 Treaty on Higher Education for Sustainable Development, and carve more sustainable pathways for all.

Professor Daniella Tilbury
President, COPERNICUS Alliance of Universities,
Dean of Sustainability, University of Gloucestershire, UK

Acknowledgements

This book is the result of a collective journey that started with a workshop in early 2012. The workshop was convened with the support of the Austrian Federal Ministry of Science, Research and Economy and sought to explore how quality assurance systems in higher education can contribute to transformations towards sustainable development. The workshop received enthusiastic support from practitioners involved in education for sustainable development around the world and later grew to include discussions on how the very concept of quality in higher education might, itself, be transformed. We all became particularly interested in figuring out together how the processes of quality assurance and improvement that were under way internationally in higher education might be integrated more effectively with the many successful innovations that had arisen from the societal call for higher education transformation including the Decade of Education for Sustainable Development, 2005–2014. We thank all of these people for their support and insights on what a book with this focus might look like.

In the course of the research and writing we benefited from the support of the United Nations University – Institute for the Advanced Study of Sustainability (UNU-IAS), particularly by encouraging participation of the Regional Centres of Expertise on Education for Sustainable Development (RCEs), many of which are represented by the authors and the chapter reviewers.

We wish to acknowledge the involvement of the International Association of Universities (IAU) and the COPERNICUS Alliance – European Network on Higher Education for Sustainable Development – for their active support for the book and assistance in reviewing the manuscript. The staff of Palgrave Macmillan have been critical in finalizing the book and we would like to thank them for their constructive assistance in getting the manuscript ready for publication. Most importantly, we would like to thank our authors and their institutions for providing the insights and case studies of how they have been transforming quality for sustainable development in universities and higher education around the world.

Addressing the challenges of social, cultural, economic and environmental sustainability that will face us all over the coming decades will require productive collaboration and the leverage of insights and experience across many disciplines, contexts and roles. This book is an example of how this can be done. It has, indeed, been a fine team effort. Finally, we would like to thank the readers of this publication, hoping that we will be able to exchange reflections and experiences and, collectively, contribute to the transformation of higher education and of society.

Zinaida Fadeeva, Laima Galkute,
Clemens Mader and Geoff Scott

Notes on Contributors

Abd Malik Abd Aziz is Assistant Director at the Higher Education Leadership Academy, Ministry of Higher Education, Malaysia. His research focuses on Sustainable Development, Education for Sustainable Development and Sustainable Livelihood Approaches. He is the co-author of *Sustainability Indicators: A CGSS Approach to Sustainability Assessment and Assistance. Sustainability Practices in Higher Education Institutions* (2013). His current research interest is in vulnerability and adaption of food security among the hardcore poor.

Andrea Bernhard is Head of the Welcome Centre at the Graz University of Technology, Austria. She studied Education and European Ethnology at the University of Graz (Austria and the University of Tampere, Finland). Between 2006 and 2008 she worked for the Austrian Accreditation Council (Ministry of Science and Research). Between 2008 and 2011 she worked as a research fellow in the Department of Educational Sciences at the University of Graz. After finishing her doctoral thesis titled 'Quality Assurance in an International Higher Education Area' (2011) she was policy advisor for the Bologna Process, Higher Education in Europe and Teaching at Universities Austria in Vienna, Austria (2012–2014). Her main research interest focuses on higher education issues, such as quality assurance, international comparisons, the Bologna process and learning outcomes. Furthermore, she investigates in the fields of adult and further education, new forms of knowledge and inter- and transdisciplinary research practices.

Katja Brundiers is the Community–University Liaison for the School of Sustainability at Arizona State University, where she develops and supports student-centred research on sustainability problem-solving in collaboration with community partners from not-for-profit, civil society, public and private sectors. To enhance students' success in such collaborations, Katja teaches an introductory course for undergraduate students on practical skills in sustainability. Her research interests relate to how to design, implement and evaluate teaching/learning settings in higher education, which support collaboration among instructors, students and community partners aimed at developing solution options to sustainability challenges. Her recent publications include *The Role* *of Transacademic Interface Managers in Transformational Sustainability Research and Education*, presenting the tasks and capacities needed for professionals intending to support collaboration between university and community partners and an educational approach for how to train students in such roles. Katja holds a master's in Geography and Anthropology from the University of Zurich.

Arthur Lyon Dahl is President of the International Environment Forum, board chairman of the ethical business building the future, and a retired Deputy Assistant Executive Director of UNEP. He has been a consultant on sustainability indicators and assessment to the World Bank, World Economic Forum, UNESCO and UNEP. He was a visiting professor, University of Brighton, researching values-based indicators of education for sustainable development, and leads a workgroup of the Partnership on Education and Research for Responsible Living (PERL), adapting these indicators for secondary schools. He teaches sustainable development in university advanced studies programmes and online courses. A biologist specializing in small islands and coral reefs, he organized the Pacific Regional Environment Programme and co-ordinated the UN System-Wide Earthwatch. His books include *Unless and Until: A Baha'i Focus on the Environment* (1990) and *The Eco Principle: Ecology and Economics in Symbiosis* (1996).

Mario Diethart works as a research associate at the RCE Graz-Styria at University of Graz, Austria. He graduated in Environmental System Sciences with an emphasis on geography at University of Graz. He has experience in various EU projects in the field of education for sustainable development with a focus on innovative teaching methods and he is also involved in international e-learning courses as a teacher.

Thomas Drage is a doctoral candidate and university assistant at the University of Graz, where he teaches education for sustainable development in the Department of Geography and Regional Sciences. He graduated in studies of Environmental System Sciences focusing on human geography at the University of Graz with his thesis being 'E-mobility as a part of sustainable development on the example of Pedelecs in Graz'. In his PhD project he studies how to foster health and well-being in the context of housing and urban planning. He worked in the field of mobility at FGM-AMOR Austrian Mobility Research, and in the department for citizen participation of the city of Graz.

Zinaida Fadeeva is Senior Specialist for Strategy and Policy at the United Nations University Institute for the Advanced Studies of Sustainability

(UNU-IAS). She has been leading research activities of the ESD team of UNU-IAS since 2003. She has published widely on the issues of ESD, change, public–private partnerships and higher education. She works with many international sustainability processes related to education and capacity development including the UN Interagency Committee for DESD, UN Alliance on Education, Training and Awareness Building of the UNFCCC, and the UN Alliance for the 10-Year Framework of Programmes on Sustainable Consumption and Production.

Laima Galkute is Associated Professor at Vilnius University, Lithuania, leading courses on Strategies for Sustainable Development and Social Innovation Projects. She has been a Coordinator of the RCE Lithuania network since 2013. Her interests focus on ESD and strategic management for sustainable development. She is a member of national education development bodies, contributes to ESD-related work of international organizations like UNECE and UNESCO and collaborates in many ESD projects.

Salfarina Abdul Gapor is Associate Professor at the School of Built Environment, University College of Technology Sarawak, Malaysia. She was also the Project Coordinator for the Sustainable Livelihood Approaches community engagement programme at Albukhary International University. Her main research interests are on sustainability issues, including Education for Sustainable Development and Sustainability Sciences.

Ellen Hazelkorn holds a joint appointment as Director, Higher Education Policy Research Unit (HEPRU), Dublin Institute of Technology, Ireland, and Policy Advisor to the Higher Education Authority (HEA). She is also President of EAIR (European Higher Education Society) and Chairperson of the EU Expert Group on Science Education (2014). Ellen has held positions as Vice President of Research and Enterprise, and Dean of the Graduate Research School (2008–2014), and Vice President and Founding Dean of the Faculty of Applied Arts, Dublin Institute of Technology (1995–2008). She works as a consultant/specialist with international organizations and universities, has been/is a member of government/international review teams and boards, and regularly undertakes higher education strategic evaluations and peer-review assessments for European and national research/scientific councils and universities. Ellen is a member of various editorial advisory boards. She was awarded a BA and PhD from the University of Wisconsin, Madison, and the University of Kent, UK, respectively. She has authored/co-authored numerous peer-reviewed articles, policy briefs, books and book chapters, including *Developing Research in New Institutions* (2005), and *Rankings and the Reshaping of Higher Education: The Battle for World-Class Excellence* (2011; forthcoming 2nd edn 2015). She is a regular contributor to the *Chronicle of Higher Education*'s Worldwise blog (http://chronicle.com/blogs/worldwise/).

Daniel J. Lang is Professor of Transdisciplinary Sustainability Research, co-director of the Institute of Ethics and Transdisciplinary Sustainability Research and Dean of the Faculty of Sustainability at Leuphana University Lüneburg. His main research and teaching interests are in the fields of theoretical, methodological and procedural fundaments of sustainability sciences, sustainable resource management, urban and regional transitions and transdisciplinary case study teaching. He graduated in Environmental Sciences at ETH Zurich and worked as senior researcher at the Institute for Environmental Sciences, Natural and Social Science Interface, ETH Zurich. In 2008 he stayed as research affiliate at the Center for Industrial Ecology, Yale University.

Clemens Mader is Post-Doctoral Fellow at the UNESCO Chair in Higher Education for Sustainable Development at Leuphana University of Lüneburg, Germany and at the Sustainability Team of University of Zurich (UZH), Switzerland. His transdisciplinary research and lectures focus on the assessment of transformative sustainability processes in the nexus of civil society, policy, education and research. Clemens is President-elect (2015–2016) of the COPERNICUS Alliance, the European Network on Higher Education for Sustainable Development and is a member of the editorial advisory board of Emerald's *Sustainability Accounting Management and Policy Journal*. Through international research and teaching affiliations he held positions in Austria, Germany, Japan, Serbia and Switzerland.

Marlene Mader works as a research associate at the RCE Graz-Styria at University of Graz, Austria, and at the UNESCO Chair in Higher Education for Sustainable Development at Leuphana University Lüneburg, Germany. She graduated in Environmental System Sciences with an emphasis on human geography at the University of Graz. Her research focuses on higher education for sustainable development, capacity building for sustainable development and scaling of sustainability knowledge. Marlene is responsible for the coordination of EU projects in the field of education for sustainable development in the interface of higher education and society.

Kathrin Maier is Quality Manager at the University of Graz. She studied Educational Science and worked for many years in the field of continuing education and quality enhancement. Her current field of work is the implementation, communication and maintenance of a quality system, to frame quality management as a tool for quality culture, and coordination of the institutional audit process.

Steven Mannell is Founding Director of Dalhousie University's College of Sustainability, and developed its innovative Environment, Sustainability & Society undergraduate programme. Recognized as one of '25 World Good Practices in Education for Sustainable Development' by UNESCO in 2009, the College approach resonates with audiences and institutions around the world. Steven co-teaches the Introduction to Environment, Sustainability & Society class, combining his broad knowledge of the history and technology of architecture, design and the built environment with his passion for the environment, interdisciplinary research, and exploring the 'wicked problems' facing contemporary society. He is a practising architect and professor of architecture, teaching in design, technology, history and practice; his publications consider the role of water in society, and advocate the reassessment and conservation of our modern built heritage. Other research includes design-build studies of spatial improvisation in lightweight building techniques of the 20th century, and community-based sustainable building.

Jonas Meyer is a research associate and lecturer at the RCE Graz-Styria at University of Graz, Austria. He graduated in Sustainable Urban and Regional Development at the University of Graz with a master's thesis on the attractiveness of native regions as a residence for highly qualified people. Currently, he is working on his PhD project, dealing with regional knowledge potentials and its cooperative use. His research focuses also on regional learning, education for sustainable development and social entrepreneurship.

Andreas Raggautz has since 2004 been responsible for the development, implementation and daily functioning of the quality management system, internal and external performance agreements and internal reporting system at the University of Graz. His main interest is the connection between quality management and strategic steering. He is a deputy representative for Austrian universities in the general assembly of the OECD-IMHE and co-speaker of the Austrian network of university quality managers.

Christian Rammel is Assistant Professor at the Vienna University of Economics and Business. Since 2011 he has been the Head of the Regional Centre of Expertise on Education for Sustainable Development in Vienna (RCE Vienna). Between 2006 and 2008 he was the Austrian representative at the United Nations Commission on Europe (UNECE) expert committee on indicators for education for sustainable development. His most recent research has focused on quality management and 'sustainable university', learning and change in complex adaptive systems as well as resilience.

Dzulkifli Abdul Razak is Chair of Islamic Leadership, Faculty of Leadership and Management, Islamic Science University of Malaysia. He is the 14th President of the International Association of Universities (IAU), a UNESCO-affiliated organization, based in Paris. Prior to this, he was the President of Association of Southeast Asia Institutions of Higher Learning (ASAIHL) (2007–2008). He served as the 5th Vice-Chancellor of *Universiti Sains Malaysia* (USM) and held the office from 2000 to 2011. Since then, he has been Founding Vice-Chancellor of the Albukhary International University (2011–2013). His other involvements internationally include membership of the Asia-Europe Meeting (ASEM) – Advisory Education Hub Committee since 2007, and the Executive Council of the Association of Commonwealth Universities (2006–2011). He currently serves as an Honorary Professor at the University of Nottingham. He is the Founding Convenor of one of the seven pioneering Regional Centres of Expertise on Education for Sustainable Development, based in Penang, Malaysia in 2005.

Paul Rowland is President of Higher Education Consulting of Highlands Ranch, Colorado, USA. Previously he was executive director of AASHE, and dean of education at both the University of Idaho and the University of Montana. He was also director of academic assessment and director of the Center for Environmental Sciences and Education at Northern Arizona University. He has written extensively about science and environmental education and higher education effectiveness.

Zainal Abidin Sanusi is Deputy Director, Centre for Leadership Training at Higher Education Leadership Academy of the Ministry of Education Malaysia. He is seconded to the Ministry from the Department of Political Science, School of Social Sciences, Universiti Sains Malaysia (USM). Zainal was the first Coordinator of Regional Centre of Expertise on Education for Sustainable Development, Penang and Deputy Director, Centre for Global Sustainability Studies, Universiti Sains Malaysia. His area of specialization is international political economy with an interest in governance for sustainable development. He has worked on several international research grants on sustainable development-related issues in a number of international agencies such as the Japan Foundation, the

Sumitomo Foundation, UNESCO and the Ministry of Environment Japan. He also did a consultation job at the Institute for Global Environmental Strategies, Hayama, Japan on various environment and development projects of the institute.

Emma Savage is a research associate at the College of Sustainability (CoS), Dalhousie University. Her research focuses on the programme development and evaluation of a new undergraduate Sustainability Leadership Certificate, available to students across the university and now in the pilot phase. She is also leading the design of a developmental evaluation protocol for programme-level outcomes of the Environment, Sustainability & Society Major. Before joining the CoS, she worked as a sustainability advisor for the Environment Network, an organization in Collingwood, Ontario that promotes sustainable living through community-based programming. Emma is a member of The Natural Step Emerging Leaders Program. She holds a master's in Environment and Sustainability from the University of Western Ontario and a graduate course certificate in Strategic Sustainable Development from Blekinge Institute of Technology in Sweden.

Geoff Scott is an emeritus professor and has been leading and researching change in higher education since the early 1970s. He has been a Pro Vice-Chancellor, an Executive Director of Sustainability and a Provost. His work has been widely published, and he led the team of Daniella Tilbury, Leith Sharp and Elizabeth Deane, who produced the 2013 international report on 'Turnaround Leadership for Sustainability in Higher Education'. He is a Fellow of the Australian College of Education, a HE Quality Auditor in a number of countries and in 2007 was the recipient of the Australian HE Quality Award.

Nadine Shovakar is a policy advisor for International Relations at Universities Austria in Vienna, Austria. She has graduated from the University of Graz completing her master's in Business Administration and in Spanish Literature. During her studies she interned at the Austrian Trade Commission in Chile and worked as a language assistant in France. Since April 2008 she has worked for Universities Austria and co-authors the *Internationalisierungspanorama*, a newsletter on both local and global questions regarding internationalization. She dedicates part of her work life to teaching English as a trainer in further education programmes at the University of Vienna's 'Sprachenzentrum'. She has recently undertaken leadership training in Education for Sustainable Development (ESD), where she assesses the quality of school projects for

ESD in Mumbai, India.

Monika Urbanski is Programs Coordinator/Analyst at the Association for the Advancement of Sustainability in Higher Education (AASHE), where she conducts data analysis related to higher education sustainability, oversees project management, and coordinates AASHE publications. In her previous role at Frostburg State University, Monika has worked in institutional research and planning, with a focus on institutional and student learning assessment. She has an MBA from Frostburg State University. Her published works include the annual *AASHE Higher Education Sustainability Review*, the biannual *Campus Sustainability Staffing Survey Report*, several campus sustainability how-to guides, and STARS annual and quarterly reviews.

Oliver Vettori is Director of Programme Management and Quality Management at WU (Vienna University of Economics and Business, Austria), including the responsibilities for the central curriculum management, the teaching and learning infrastructure and the university's teaching and learning services. He is a team coordinator for the Institutional Evaluation Programme of the European University Association and is currently also serving as a member of the Steering Committee of the European Quality Assurance Forum. He works regularly as an evaluator, trainer and consultant for various international networks and institutions. As a teacher and research associate at the Institute for Organization Studies (WU), his current professional and research interests lie in the areas of organizational dynamics, interpretive patterns and meaning structures. His most recent publications include 'Dealing with Engagement Issues – An Examination of Professionals' Opinions on Stakeholder Involvement in Quality Assurance' (with Tia Loukkola, EUA 2014) and 'Finding Meaning in Higher Education: A Social Hermeneutics Approach to Higher Education Research' (with Manfred Lueger, in *Theory and Method in Higher Education Research* II, 2014).

Arnim Wiek is Associate Professor in the School of Sustainability at Arizona State University and Head of the Sustainability Transition and Intervention Research Lab. His research group conducts sustainability research on emerging technologies, urban development, resource governance, climate change, and public health in USA, Canada, different European countries, Sri Lanka, Mexico and Costa Rica. The group develops evidence-supported solutions to sustainability challenges in close collaboration with government, businesses and community groups.

Arnim Wiek holds a PhD in Environmental Sciences from the Swiss Federal Institute of Technology Zurich, and a master's in Philosophy from the Free University Berlin. He had research and teaching engagements at the Swiss Federal Institute of Technology Zurich, the University of British Columbia, Vancouver, and the University of Tokyo.

Friedrich M. Zimmermann is Professor and Chair in the Department of Geography and Regional Science, University of Graz, Austria and is Director of the RCE Graz-Styria (UN-certified Regional Centre of Expertise on Education for Sustainable Development). He was Vice-Rector for Research and Knowledge Transfer (2000–2007) and is currently sustainability commissioner at the University of Graz. He is the founding President and Advisory Board member of the COPERNICUS Alliance, the European Network on Higher Education for Sustainable Development. He had international affiliations at the University of Munich, at universities in Pennsylvania and Oregon, US, and in Croatia and Serbia. His research focuses cover sustainable urban and regional transformation processes; sustainable tourism planning and prognosis; sustainability and knowledge transfer. Besides numerous published articles in books and reviewed journals, Zimmermann is a co-editor of books and journal editor. He is working with several international and interdisciplinary research teams in projects of the UNESCO and United Nations University–IAS, the European Council, the European Science Foundation, the European Partners for the Environment, the Austrian Development Cooperation, the Austrian Academy of Science, and in EU projects, such as the EU-ENRICH Programme, EU-INTERREG, EU-ESPON, EU-TEMPUS-Programmes, EU-Comenius-Programme, EU-e-Learning Programmes, and so on.

List of Abbreviations and Acronyms

AASHE	Association for the Advancement of Sustainability in Higher Education
ABCD	Asset Building and Community Development
ACUPCC	American College and University Presidents Climate Commitment
AHELO	Assessment of Higher Education Learning Outcomes
AHSS	arts, humanities and social science
AIU	Albukhary International University
API	application programming interface
ARDE	Accountable Research Environments for Doctoral Education
ARWU	Academic Ranking of World Universities
AUA	Alternative University Appraisal
AUQA	Australian Universities Quality Agency
BFUG	Bologna Follow-up Group
BIQ	benchmark indicator questions
CDD	Community Driven Development
CGPA	cumulative grade point average
CoS	College of Sustainability
CRASP	Conference of Rectors of Academic Schools in Poland
CSDC	Campus Sustainability Data Collector
CSR	corporate social responsibility
CWTS	Centre for Science and Technology Studies
DE	developmental evaluation
DESD	Decade of Education for Sustainable Development
EBBF	European Bahá'í Business Forum
EfS	education for sustainability
EGM	emerging global model
EHEA	European Higher Education Area
EQUIS	European Quality Improvement System
ER	Education and Research
ESD	education for sustainable development
ESG	European Standards and Guidelines
ESS	Environment, Sustainability, and Society
EU	European Union
EUA	European University Association

FINHEEC	Finnish Higher Education Evaluation Council
GFC	global financial crisis
GHG	greenhouse gas
GMID	Graz Model for Integrative Development
GSF	gross square feet
GT	Gestion de Terroir
GWS	Greater Western Sydney
HCF	Humaniversity Competency Framework
HE	higher education
HEASC	Higher Education Associations Sustainability Consortium
HEEACT	Taiwan Performance Ranking of Scientific Papers for Research Universities
HEI	higher education institution
HESI	Higher Education Sustainability Initiative
IAU	International Association of Universities
IFRC	International Federation of Red Cross/Red Crescent Societies
IN	innovation
KPI	key performance indicator
LEED	Leadership in Energy and Environmental Design
LiFE	Learning in Future Environments
L&T	learning and teaching
M&E	monitoring and evaluation
MoU	Memorandum of Understanding
MQA	Malaysian Quality Accreditation Agency
NCAA	National Collegiate Athletic Association
NGO	non-governmental organization
NIS	national innovation system
NQF	National Qualification Framework
NTU	National Taiwan University Ranking
OEAD	Österreichische Austauschdienst – Austrian Agency for International Cooperation in Education and Research
OECD	Organisation for Economic Co-operation and Development
OP	operations
PAE	Planning, Administration and Engagement
PIMAUG	Environmental Institutional Programme of Guanajuato University
PPBL	Problem- and Project-Based Learning
ProSPER.Net	Promotion of Sustainability in Postgraduate Education and Research Network

QA	quality assurance
QM	quality management
R&D	research and development
RBA	Rights Based Approach
RCE	Regional Centre of Expertise on Education for Sustainable Development
SAQ	self-awareness questions
SD	sustainable development
SEI	Sustainable Endowments Institute
SLA	Sustainable Livelihood Approaches
SRAS	Sustainability Rating Assessment System
SSN	Social Safety Net
STARS	Sustainability Tracking, Assessment & Rating System
SUNY	State University of New York
SWOT	strengths, weaknesses, opportunities, threats
THE	Times Higher Education World University Ranking
THE-QS	Times Higher Education-QS World University Ranking
TILT	Tracking and Improvement System for Learning and Teaching
TIM	transacademic interface manager
UCC	University College Cork
UN	United Nations
UNESCO	United Nations Educational Scientific and Cultural Organization
uniko	Österreichische Universitätenkonferenz – Universities Austria
UNU	United Nations University
UNU-IAS	United Nations University Institute for the Advanced Studies of Sustainability
URAP	University Ranking by Academic Performance
USNWR	US News and World Report of Best College Rankings
UWS	University of Western Sydney
WoS	Thompson Reuter's Web of Science
WW2	World War 2

Structure of the Book

The book is structured around three parts, each of them highlighting a particular set of themes. Contributing authors to Part I (Chapters 2–4) demonstrate how recognition of higher education as the 'engine of development' puts the sector increasingly at the centre of discussions about societal transformation, which had earlier been, predominantly, the domain of the sustainable development discourse (and subsequently the education for sustainable development discourse) (Box 0.1). The authors discuss the evolution of HE assessment – from the impact of a focus on rankings to linking sustainability issues to the development of an institution's quality culture and the systems that underpin it and effective ways of managing the interplay between external and internal quality assurance – against the backdrop of evolving understanding of knowledge and knowledge production.

Box 0.1 Transformation, sustainability and higher education

- The calls for greater service to the society and change have been, for many years, part of the sustainable development and education for sustainable development discourses. With the numerous initiatives in curriculum innovation, research and community outreach being undertaken by different universities around the world, the calls for HE to transform itself as a system come as a relatively recent global trend. For example, the Higher Education Sustainability Initiative (HESI) for Rio+20, initiated in 2012 by a group of UN partners, calls for HEIs to address ESD by promoting 'development through both research and teaching, disseminating new knowledge and insight to their students and building their capabilities'.
- The People's Sustainability Treaty on Higher Education, another bottom-up initiative of higher education towards Rio+20, has emphasized the significance of the policies and mechanisms that would ensure alignment of the higher education practices with sustainability. The latter aspect is critically significant as it puts sustainability goals outside of the often narrow domain of practices of its promoters into the mainstream discussions of transformation of the society and, consequently, higher education.

Part II (Chapters 5–8) explores the meaning and the role of internal quality assurance and its interplay with external quality approaches in supporting HE sustainability transformation. The authors demonstrate the significance of considering the question of quality at all levels – programme, organization, national and international – while paying close attention to the interplay of the dynamics between HE development and ESD. They highlight the rich variety of quality assurance processes that are transforming higher education and how these can be used to enhance the focus and quality of ESD in different regions. Most importantly, each of the authors emphasizes the *changing meaning of quality* and the evolving role of HE as a result of recent social, economic and political developments and how all these forces are working together to help transform HEIs.

Part III (Chapters 9–12) focuses on the ways that quality systems facilitate the transformative function of higher education by supporting the development of sustainability capabilities and competences and high quality learning designs at both the course and programme level. In the context of societal transformation, educational institutions are required to cultivate both instrumental and emancipatory competences. Instrumental competences are seen as being more utilitarian whereas emancipatory competences (in some jurisdictions called 'capabilities') refer to the human ability to handle unexpected challenges promptly and effectively, learn and grow, and to reflect, interact and engage productively with others. To be effective, they are to be considered in a discussion with learning processes.

The authors analyse and identify how quality management for key areas like ESD programme design, delivery, implementation, impact and quality improvement can be linked with other strategic elements of HEIs such as leadership aspirations, goals and planning processes, as well as with staff and partnership development, the validation of learning outcomes and their assessment and how, through this, institutions can be transformed to become more change capable and focused on ESD.

1
Assessment for Transformation – Higher Education Thrives in Redefining Quality Systems
Zinaida Fadeeva, Laima Galkute, Clemens Mader and Geoff Scott

A new role for higher education in the 21st century

Higher education has a unique opportunity to provide learning for the future and help the world address the rapidly unfolding social, cultural, economic and environmental sustainability challenges of the 21st century. However, to fulfil this role at the regional, national and international levels, higher education institutions themselves have to undergo critical transformation towards sustainable development in their philosophy and practices and put in place the quality assurance systems to ensure that this transformation is consistently implemented and effective.

'Quality' is a term much used in many contexts, including in education at all levels across the world. A key objective of this book is to illuminate what is meant by this concept in the distinctive context of tertiary education, to develop a shared understanding of the need to transform it and to provide proven and feasible ways in which to achieve this. It is common to hear university educators defining 'quality' as meaning 'fitness for purpose' but the authors in this volume look not only at learning designs, research and engagement processes being 'fit for purpose' to ensure productive research, teaching, engagement projects and campus operations but also at the very fundamental purpose of 21st-century higher education itself. That is, they argue that quality should not only be defined as fitness for purpose but also as being very much about *fitness of purpose*. And from this standpoint they then go on to look at higher education's *fitness for transformation* as a key driver for sustainable development and societal improvement.

Most higher education systems around the world have put in place a range of quality assurance, auditing and accreditation systems over the past three decades. There has been a general shift from looking at simple quality control systems to building internal capability for continuous quality assessment and improvement. Box 1.1 gives definitions of quality assurance and quality management provided by UNESCO-CEPES:

Box 1.1 Definitions

Quality assurance:

An all-embracing term referring to an ongoing, continuous process of evaluating (assessing, monitoring, guaranteeing, maintaining, and improving) the quality of a higher education system, institutions, or programmes. As a regulatory mechanism, quality assurance focuses on both accountability and improvement, providing information and judgments (not ranking) through an agreed upon and consistent process and well-established criteria.

Quality management

An aggregate of measures taken regularly at system or institutional level in order to assure the quality of higher education with an emphasis on improving quality as a whole. As a generic term, it covers all activities that ensure fulfilment of the quality policy and the quality objectives and responsibilities and implements them through quality planning, quality control, quality assurance, and quality improvement mechanisms.
Quality assurance is often considered as a part of the quality management of higher education, while sometimes the two terms are used synonymously.
Source: UNESCO-CEPES, 2007.

This book seeks to link key developments and experience in higher education quality assurance and improvement systems and what we have learnt about effective change management in HEIs with the need to transform our universities and colleges to give greater focus towards becoming more sustainable and resilient societies in all their activities. It gives particular attention to the:

- transformation of higher education towards fostering sustainable development in society;
- various ways in which quality assurance and improvement processes are defined and used, and are subject to differences in societal context;
- interplay of quality processes at different levels: international, national, organizational and in HEI's educational programmes;

- role of good practice in quality and change management in supporting transformation of HEIs;
- role of quality management in supporting the development of the capabilities and competences of graduates and the capabilities of HEIs required towards a just and resilient society.

Contrary to the dominant perception of quality management as a tool for compliance (or, in the case of higher education ranking systems, as a marketing and differentiation tool), the authors in this book present quality assurance as instrument for transformation that can help reshape the strategic, cultural and political dimensions of HE life. At this turnaround moment for HEIs (see Barber et al., 2013), our authors identify new notions of quality that leverage diversity and recognize the dynamic and complex nature of institutional transformation towards a more embedded focus on education for sustainable development (ESD).

Quality assurance as an instrument for transformation

A number of chapters argue that, if used as a tool not just for compliance and quality control but also for continuous improvement, both external and internal quality management systems can be powerful instruments for the transformation of both universities and those who populate them, whilst simultaneously fostering the diversity necessary for institutional and social sustainability. As Ellen Hazelkorn in her chapter emphasises, the application of knowledge is widely acknowledged as being the source of social, economic and political power.

Strong links are identified in many chapters (for example in the chapters by Mader, Vettori & Rammel, Shovakar & Bernhard, Urbanski & Rowland, Zimmermann et al.) between building a quality-focused, evidence-based, change-capable culture in our universities on the one hand and, on the other hand, what distinguishes change-capable, resilient, adaptable graduates, organizations and societies.

It is argued, for example, that effective strategies for managing quality, successful change management for ESD and effective approaches to sustainable development in society all adopt a 'whole of institution', 'systems approach'. In doing this they use not only internal and external quality systems as levers to motivate and support engagement in the transformation of our higher education institutions towards a more systematic focus in their core activities on the four pillars of sustainability but they also use a combination of top-down and bottom-up approaches (see, for example, the chapters by Galkute and Shovakar & Bernhard) via a process of 'steered engagement' (as outlined in the chapters by Scott and Vettori & Rammel). That is, they look at the ecology of the total of the university in exactly the same way that we need to understand and take into account the total ecology of our world in order to foster effective transformation and resilience. Other chapters (e.g. Galkute, Mader and Zimmermann et al.) suggest ways in which university roles and responsibilities and the operating principles that underpin them can be reshaped to give greater focus to quality and sustainability.

Contributors and contributions

The chapters which follow are written by practising educators, researchers, innovators and leaders from a wide variety of higher education contexts ranging from Europe and North America to the Asia-Pacific. Collectively, they present a wide range of perspectives and the key lessons learnt about forward-looking approaches to assuring and improving quality and achieving successful change management for ESD in an extensive range of operating environments.

Two overarching themes underpin the perspectives and strategies identified in the book for ensuring the quality and successful implementation of initiatives aimed at embedding ESD into the core business of our higher education institutions. They are: 'Good ideas with no ideas on how to implement them are wasted ideas and Change doesn't just happen but must be led, and deftly' (Scott et al., 2012, p. 8). Whilst much has been written about *what* should change in the area of transformative higher education much less has been written about *how* to ensure that these desired transformative changes actually get put successfully and sustainably into practice, with consistent quality. And it is around this issue and the role of successful ESD change management and leadership in higher education that this book turns.

The need for shared meaning and common terminology is one of the recurring themes in the contributions. A number of authors (e.g. those from Australia, Asia and North America) note the need for us to make sure that we are not talking at cross purposes when discussing transformation towards sustainability in higher education. They not always use terms and concepts like *sustainability, sustainable development, quality, quality assurance, quality audit, quality improvement, assessment* and *evaluation* with the same meaning. In spite of this broad diversity, the authors show a common understanding in considering quality assurance as instrument for transformation including at the level of working with competences, values and areas of learning.

A successful quality and standards framework for ESD is identified and associated terms are defined by Brundiers et al. in Chapter 10 of the book. This framework emphasizes that it is the total university experience that engages and retains students in productive learning, not just what happens in the traditional classroom (Scott, Chapter 11), along with a wide range of key tracking measures and systems that can be used to ensure that what is planned, using such a framework, is actually being put successfully and consistently into practice and that key areas for improvement are promptly identified and addressed. The chapters by Gapor et al., Urbanski & Rowland and Brundiers et al. provide additional details on this issue.

A variety of *successful options for learning for sustainable development* is another feature of the book. Many proven and productive ways to help students learn about ESD and ways to ensure that the assessment of what is learnt is relevant, valid, helpful and reliable are outlined in the book. Case studies of effective learning methods for ESD include, in Australia, the use of the campus as a living laboratory for learning about and researching ESD (Scott); in the US and Canada the use of Problem- and Project-Based Learning (PPBL) (Brundiers et al.) and the involvement of students in using the STARS tracking system to advance campus sustainability (Urbanski & Rowland); in Malaysia the use of the Sustainable Livelihood Approaches (SLA) to learning (Gapor et al.); and in Austria a range of student engagement initiatives including BioTechMed-Graz and the 'Sustainicum' (Shovakar & Bernhard). The chapter by Dahl gives specific focus to the issue of what values should be developed in our graduates.

A group of authors (Galkute, Mader and Shovakar & Bernhard) present examples of organizational and mutual learning in a course of HEIs' alignments of organizational strategy development and change management as well as attempts to balance organizational learning and development of students' competences together with society.

Interestingly, the authors refer to *a range of opportunities for networked learning* which can help provide proven solutions to key quality improvement priorities for ESD at HEIs. For example, the chapters by Zimmermann et al. and Mader note that the development of some 129 Regional Centres of Expertise in ESD by the UN University has great potential to give focus to this work and provides an ideal international and regional multi-stakeholder learning network on sustainability ideas and solutions. The chapter by Urbanski & Rowland, as well as the one by Shovakar & Bernhard, show the role of networking in integrating principles of ESD into universities' operations and assessment systems at the national level in the USA and Austria, respectively.

Rethinking the role of knowledge

Current interest in rethinking and redesigning the focus of our HEIs and systems of knowledge creation to give increased focus to ESD are a reaction to dramatic changes in our social, political, financial and environmental context over the past three decades. They include experiencing a series of financial and environmental crises which have led to the emergence of what some have called a 'civilizational community of fate' (Beck and Cronin, 2006, p. 13) in which the actions of different communities of practice, professions and groups interconnect and influence each other at the local and global level.

The new context is seen as requiring much more two-way collaboration between 'knowledge professionals' on the one hand and, on the other hand, groups beyond the university in order to frame problems, and identify, test, refine and scale-up solutions effectively. This approach, it is argued, helps ensure that such solutions are situated, feasible, relevant and context-specific. The emerging importance of understanding context and appreciating other forms of knowledge manifests itself in greater acknowledgement of traditional understandings of how to foster sustainable development that go beyond the dominant 'modern' knowledge traditions.

For example, modern methods of knowledge production and validation are fundamentally different from the approaches of 'traditional knowledge' systems which are typically non-dualistic, dynamic, informal, sacred, spiritual, time related and not linear in nature. Integration of modern and traditional knowledge creation systems that enable us to go beyond simple utilitarian use of traditional practices or their protection signifies one of many frontiers that need to be addressed in a new approach to higher education for development (Payyappallimana et al., 2013).

Having to manage the complex mix of risks and rapid and unpredictable changes in society, the economy and politics, along with the recognition of other knowledge systems and knowledge stakeholders, has important implications for universities and for the entire higher education system. In this regard the chapter by Ellen Hazelkorn (Chapter 2) presents a compelling picture of the evolution in knowledge development. She emphasizes that application of knowledge is widely acknowledged as being the source of social, economic and political power. With the deepening linkages between higher education and society now under way around the world the focus is increasingly on shifting from 'curiosity-oriented' research to 'socially robust, collaborative and interdisciplinary' knowledge development whilst ensuring greater social and public accountability of HEIs in ways that are consistent with Agenda 21 (Box 1.2).

> Box 1.2 Science for sustainable development
>
> The sciences are playing an important role in linking the fundamental significance of the Earth system as life support to appropriate strategies for development which build on its continued functioning. The sciences should continue to play an increasing role in providing for an improvement in the efficiency of resource utilization and in finding new development practices, resources, and alternatives.
> *Source*: United Nations, 1992, chapter 35.

Changing society by changing the capabilities and competences given focus in higher education

The capability of our graduates to contribute creatively and constructively to the (sustainable) development of our nations is of critical importance in the current context. As Scott et al. (2008) note:

> Whereas being competent is about delivery of specific tasks in relatively predictable circumstances, capability is more about responsiveness, creativity, contingent thinking and growth in relatively uncertain ones. What distinguishes the most effective [performers]...is their capability – in particular their emotional intelligence...and a distinctive, contingent capacity to work with and figure out what is going on in troubling situations, to determine which of the hundreds of problems and unexpected situations they encounter each week are worth attending to and which are not, and then the ability to identify and trace out the consequences of potentially relevant ways of responding to the ones they decide need to be addressed.

One of the crucial dimensions of capability highlighted by Sen (1999) relates not to the more (or less) narrow attributes of what learners acquire but to the idea of choice. In the context of ESD, the choices have to move beyond the limited notion of personal well-being to notions of social responsibility and ultimately, justice. This brings us to the critical role which values play in the notions of capability and competence (Chapter 9 by Dahl).

In this regard, the capabilities referred to above are closer to what others refer to as competences. As Rychen (2004) notes, these entail: 'a combination of interrelated cognitive and practical skills, knowledge (including tacit knowledge), motivation, values and ethics, attitudes, emotions, and other social and behavioural components that together can be mobilized for effective action in particular context'. In the field of education, the focus on developing relevant graduate capabilities and competences reflects a significant policy shift away from concentrating on the quality of the inputs to our HEIs to the quality of their outcomes and impact (Tiana, 2004). This, in turn, has been reflected in a shift from teacher-centred to student-centred approaches to learning, with a focus on ensuring not only close linkages between teaching, learning and assessment but also between assessment and the development of those capabilities and competences necessary to build a sustainable future. Close mutual dependence of competences and capability with learning (and assessment as learning) processes becomes significant also for the reason that lack of such interlinkages reduces competences to a mere 'checklist' of categories serving rhetoric rather than a desired outcome.

Furthermore, HEIs need to not only educate people with the capabilities and competences necessary for effective practice in the current context but also the leaders of change for tomorrow, people who can help shape a better future, who can work collaboratively with business and the institutions of civil society to develop and implement solutions that will effectively address the challenges of social, cultural, economic and environmental sustainability. In this perspective HEIs have a 'twofold' role in leading societal change: by preparing students for professional careers and by providing innovation and 'knowledge translation' to a variety of stakeholders and to society at large. These two processes can be interrelated by each university or college's management taking a whole institution approach (this issue is explored in a number of chapters, including in Part One of the book and in Chapter 11).

As a social construct, HEIs tend to reflect the predominant features of the social organization, knowledge structures, and the attitudes and values of the country in which they are located. For example, in the industrialized world higher educational systems tend to be increasingly designed to stimulate economic growth and competiveness. However, it is argued that what is now needed to ensure sustainable development is a transition towards educational systems which substantially empower learners by promoting a new culture of citizenship, social inclusion and participation and by developing futures thinking in our graduates, including a capacity to foresee change, develop creative strategies and choose relevant development paths. In this perspective it also becomes essential to identify, proactively, learning outcomes and the capabilities and competences which are important both for professional careers as well as the initiation of social change. With creativity, innovation, entrepreneurship emerging as key areas for development in a broad variety of professional fields, we recognize a shift from a 'creative class' (Florida, 2002) towards 'creative society'.

A number of the book's chapters look at this revised focus for higher education – at developing universities and colleges which go beyond producing only 'work-ready' graduates for today's markets and needs. They give emphasis to educating learners for tomorrow's needs: people who are sustainability literate, change implementation savvy, inventive, creative, entrepreneurial, culturally situated and ethically robust. They argue for HEIs which aspire to educate graduates who can come to a considered position on the little-questioned, tacit assumptions currently driving the 21st-century development agenda – assumptions like growth is good, consumption is happiness, ICT is the answer, and globalization is great. And, as a minimum, some HEIs hope that even in the rapidly changing circumstances of markets and life in the 21st century such capabilities will be able not only to sustain themselves but those around them.

The distinctive attributes identified in studies of change-capable, productive and resilient leaders, including recently completed research with 188 experienced turnaround leaders for sustainability in higher education across the world (Scott et al., 2012) and parallel studies of successful early career graduates align closely with the attributes of the most change-capable HEIs and, indeed, the most resilient, harmonious, sustainable and productive societies (Chapter 11 by Scott). They all exemplify high levels of personal and interpersonal emotional intelligence, including the ability to remain calm when things go wrong, to tolerate ambiguity, behave ethically and transparently, to practise what they preach, to listen to all the key players, to work productively with diversity, to link and leverage input from all involved, work towards a common cause, and they all apply a creative, contingent, diagnostic intelligence as they continuously seek to both shape and negotiate a continually shifting context.

Cultivating capabilities and competences for sustainable development requires relevant learning processes and environments, not only to develop the capabilities that count but also to ensure that they are assessed validly. The chapter by Brundiers and colleagues (Chapter 10) outlines experience in three different sustainability study programmes carried out in Canada, Germany and the USA. They discuss problem- and project-based learning as key tools for building key capabilities and competences in sustainability, including the combination of systems anticipatory, normative and strategic competences as well as interpersonal capabilities that are so necessary for effective practice in the current context. The course-level evaluation framework and especially the developmental evaluation framework used at the level of an entire study programme which they discuss provide important instruments for supporting innovation and the *continuous* learning of university teachers (Box 1.3).

Box 1.3 ESD competences of educators

The ESD competences of educators were particularly emphasized by the United Nations Economic Commission for Europe (UNECE, 2011). The following groups of competences were defined as the most important for educational transformation: a holistic approach (integrative thinking, inclusivity, dealing with complexities); envisioning change (learning from the past, inspiring engagement in the present, exploring alternative futures); achieving transformation (transformation of the role of educator, transformative approaches to teaching and learning, transformation of the education system as a whole). It is also emphasized that a 'whole-institution approach' should be adopted for the continuing professional development of educators in their workplace. Educators will best develop the competences when the culture and management of the entire organization is supportive of sustainable development.

Source: UNECE, 2011.

In Chapter 9 Arthur Dahl gives specific focus to the issue of what values should be cultivated in graduates and how values are intimately linked to graduate capability and competence. In identifying sustainability as an ethical concept, Dahl points out that sustainability values can be expressed differently in a variety of contexts. He notes: 'Values, beliefs and ethics are a key driver for successful education for sustainability. Values make it possible to judge behaviour that benefits society.' His research aligns with the findings of many earlier studies of successful early career graduates identified by Geoff Scott in Chapter 11, specifically a distinctive combination of personal attributes (including the ability to remain calm when things go awry, to be able to tolerate ambiguity, humility, authenticity, trustworthiness, integrity, empowerment and a commitment to justice and equity) and interpersonal attributes (including the ability to work productively with diversity, being able to listen before acting, an ability to lead through influence as well as a respect for the environment, social harmony and community). This focus on values highlights the importance of recognizing that there is a profound difference between change (something becoming or being made different) and progress (a conclusion that this is in a positive direction). Quite simply, change is a value-laden, learning (and unlearning) process for all concerned.

Adopting a whole institution approach to change

Effective impact on societal development requires, first of all, strong commitment on the side of HEIs. This means that policy, values, leadership, core activities and cooperation with key stakeholders are aligned and oriented towards institutional priorities. In the context of this book it would also mean that a focus on social, cultural, economic and environmental sustainability is embedded in the research, teaching, operations and engagement activities of the HEI. Extensive research on effective strategic change in higher education has found that what works best is a process of 'steered engagement' in which overall quality parameters and strategic development priorities are set and then local institutions, faculties and groups identify the best way to action them given local histories, resources and needs is becoming increasingly common (Fullan and Scott, 2009, pp. 85–88).

Various contributors to the book pick up these different aspects of system alignment and explore how a whole-of-system approach might best be enacted. The constructive engagement of staff as well as students becomes particularly important as it is they who will (or will not) ensure that any desired ESD innovation is actually taken up and put into practice at the organizational and local level. In Chapter 3 Vettori & Rammel highlight how engagement with the university community needs to be concerned with individual and collective values, everyday actions and interactions, and how, eventually, from this process a collective commitment to achieving the desired ESD innovations can be achieved. In Chapter 11, Geoff Scott provides a review of research on effective change leadership of sustainability in higher education and identifies a set of strategies found to be effective in transforming the entire university to take a more systematic, linked and leveraged quality assured approach. It is important to note that staff engagement is intimately tied up with what motivates people to change (i.e. learn) and is always driven by each individual deciding that what is proposed in a proposed change (e.g. a call to get involved in an ESD initiative) is relevant, desirable, clear and, most importantly, feasible (achievable and deliverable).

Gapor et al. (Chapter 12) identify how the transformative vision of the Albukhary International University (AIU), located in Malaysia, is to contribute to the elimination of poverty through the implementation of a sustainable livelihood approach that cuts across all aspects of university activity. They note how AIU has set aligned policy directions for staff and student recruitment, and how it has made ESD the core focus of the institution's curriculum, research and community engagement. Interestingly, as the ultimate goal of all the AIU's activities has become contribution to the community, including in the region immediately surrounding the university, the traditional distinction between research, education and community outreach has disappeared.

Ensuring that there is an aligned monitoring and improvement system for ESD is another aspect of taking a whole-of-system approach. A good example is the Sustainability Tracking, Assessment & Rating System (STARS) discussed by Urbanski & Rowland in Chapter 8: an initiative of the US-based Association for the Advancement of Sustainability in Higher Education (AASHE). STARS initially operated just in North American HEIs but more recently it has been taken up elsewhere. It focuses on providing a framework for HEIs to measure and self-report on their sustainability performance. The STARS programme not only covers the major areas of HEIs activities – Education & Research, Operations, Planning, Administration, Engagement and Innovation – but it also assists participating institutions to benchmark the quality of what they are doing and identify proven ways of *continuously* addressing their quality improvement priorities.

Relevant conditions for quality assurance as well as improvement of the academic standards in the area of ESD requires not only the focused, collegial leadership of institutions' administration but also support and involvement from the entire HEI community leading to a quality culture (see Chapters 8, 11 and 12 by Urbanski & Rowland, Scott and Gapor et al., respectively). More broadly, a number of chapters highlight how important it is for HEIs to develop reciprocal, participative relationships with society. Other chapters (e.g. Chapter 7 by Zimmermann et al.) suggest ways in which university roles and responsibilities and the operating principles that underpin them can be reshaped to give greater focus to quality and sustainability in a systematic way. In Chapter 4, Mader points out a need for an iterative and integrative process where, through inclusive assessment, quality systems can contribute to translation of sustainability ideas between HEIs and the society to be put successfully into practice.

The book shows how the fundamental principles and strategies for developing a sustainable world are closely related to the successful approaches now under way in higher education to manage change and assure and improve the quality of the core activities of HEIs. Motivation for change (that is, wanting to transform our HEIs towards a more integrated and embedded focus on sustainability in all their activities) is closely related to both value clarification and value judgement. This is because it is judgements of value (evaluation) that lead individuals to decide if a proposed ESD change is relevant, desirable, clear and feasible. If it is concluded that this is not the case then one will not engage or stick with learning how to make it happen in practice. In all of the research on ESD change and leadership undertaken to date (see Scott et al., 2012) it has been repeatedly found that change-capable, resilient and sustainable organizations and societies don't just happen, they are built by leaders who model the capabilities identified in Chapter 11 and, through their modelling, these leaders build a change-capable and resilient culture.

Achieving all of the interlaced changes necessary to transform HEIs towards sustainability can pose a daunting agenda. However, if each initiative is tracked during implementation and clear evidence of its positive impact on those concerned is gathered then there is much greater likelihood that those who are sceptical about ESD initiatives will be won over, especially if they are provided with evidence of successful, satisfying and productive initiatives in operating contexts and fields of education similar to their own.

The importance of making sure there is an aligned support, incentive and governance system and capable staff to deliver ESD initiatives (Mader et al., 2013), as well as robust and valid tools to assess learning and to track and improve their quality, has already been noted.

Many of our authors identify a wide range of challenges to be addressed if sustainability initiatives are to be successfully embedded in our HEIs' core business. These challenges, as Hazelkorn notes in Chapter 2, include the counterproductive role of rankings in the reconstruction of knowledge in 'the age of austerity'. Some of the other authors note the challenge of seeking to introduce interdisciplinary programmes in institutions whose structure and incentives are more mono-disciplinary; the challenge of working out ways to engage uninterested senior leaders and disengaged local staff; of achieving better alignment between resources, incentives, performance management and staff development and the objective of embedding ESD in the core activities of our tertiary institutions.

Other authors look to the issues in a vocationally focused policy environment of giving greater emphasis in higher education learning and assessment to the development of people who can help shape ethically robust societies, a culture of citizenship, social inclusion and participation as well as being able to work productively with diversity, think proactively, foresee change, create new solutions and to choose relevant development paths. A wide range of currently used strategies for assessing students' development of these capabilities is provided in a number of chapters, including those by Gapor et al. and Scott (Chapters 12 and 11).

Strategic development of competences

As noted earlier there is a tendency to give too much focus to what should change but much less attention to how exactly a good idea for ESD innovation can be taken and actually put into practice effectively, consistently and sustainably. This finding identifies an important opportunity for our institutions of higher education and those who teach in them to show their students how to manage change successfully by modelling to them effective approaches to dealing with and adapting to the distinctive issues of sustainability facing their regions as well as, more broadly, becoming beacons and exemplars of sustainable development and learning, seeking to 'practise what they preach' and to act as living laboratories for learning about, researching and engaging collaboratively with their surrounding communities in joint projects to address the four interlaced pillars of sustainable development that affect them. This is in recognition of Spohn's (2003) observation that people are more likely to act their way into new ways of thinking than think their way into new ways of acting.

A number of contributors to the book identify ways of addressing implementation challenges like those identified above using a whole institution approach. They identify a wide range of practical change, quality, leadership and improvement strategies including: educating sustainability change agents; creating a participative culture around sustainability in higher education; and showing how to take a systematic approach to university–community partnerships in sustainable development. They provide case studies of how to use effective external and internal quality systems to foster productive change in ESD and outline practical ways of implementing and using them productively, as well as suggesting ways to validly assess ESD in our universities and how to comprehensively and productively track and improve sustainability initiatives.

As noted earlier, strong links are identified in many chapters (for example in those by Vettori & Rammel, Mader, Shovakar & Bernhard, Urbanski & Rowland, Zimmermann et al. and Galkute) between building a quality-focused culture in our universities on the one hand and, on the other hand, what distinguishes change-capable, resilient, adaptable graduates, organizations and societies (Fullan and Scott, 2009).

Quality management systems for transformation

Beyond ranking

Over the past three decades the use of global quality ranking systems for HEIs has become increasingly popular. However, a range of concerns have been expressed about such systems. These include:

- the use of simplistic indicators to measure complex phenomena;
- the negative influence of taking a single discipline, single institution and 'modern' focus in many ranking systems when what is needed is more focus on transdisciplinarity; and
- taking transnational perspectives located in collaborative action research and learning about how to address key local, national and international issues of sustainability.

In Chapter 2 Hazelkorn takes up this theme and observes that ranking universities and colleges stands in direct contradiction to the required ways in which HEIs need to transform in order to remain relevant to the needs of the 21st century. It is noted that ranking privileges some disciplines over others, gives limited attention to the impact of research for the benefit of society, fails to take into account the diversity and complex nature of HEIs in all parts of the world, and that the ranking indicators 'neither reflect contemporary understanding of the spectrum of knowledge production nor measure what public policy seeks to achieve' (Hazelkorn). Still, probably owing to their simplicity and appearance of scientific rigour supporting their development, ranking remains an influential tool among higher education stakeholders leading to several undesirable consequences including values attached to the narrow fields of knowledge, loss of focus on the local and unique, or lack of motivation for engagement with newer and 'riskier' social goals and aspirations.

However, with the failure of rankings to recognize the diversity and complex nature of various HEIs, more sophisticated and transformative assessment systems, with characteristics described in the following section, have begun to emerge. These systems have emerged as a result of complex interactions, and negotiations, at different levels, including interplay between internal and external quality assessments with the later influenced by international and national processes.

Exploring the interplay between external and internal quality assurance processes

While national approaches to external quality assurance vary, there are three main modal forms (Dill and Beerkens, 2010): the central regulation of quality assurance by the state via educational ministries setting standards and accreditation/audit processes for HEIs; market regulation; and professional self-regulation. The last two can be considered as an alternative to the state approach for assuring academic quality. These modalities operate in different combinations reflecting national circumstances and public interest.

With respect to the role of HEIs in transforming society, it is important to explore the *internal mechanisms for quality assurance in relation to external approaches* as well as *factors that have an impact on their interplay*. These factors may include, in various combinations, the level of autonomy the higher education institutions have, their strategies which differ from national development strategies, stakeholder influence, and the requirements of professional accreditation bodies.

Within the European Higher Education Area (EHEA), which currently involves 47 countries, there are some commonly agreed Standards and Guidelines for Quality Assurance (EHEA, 2005) which define the principles and make recommendations concerning internal and external quality assurance. Within these parameters countries are free to choose priorities and management approaches according to their individual situation. It should be noted that the focus on quality within the EHEA is closely linked to the various missions of higher education, ranging from teaching and research to community service and engagement in assisting social cohesion and cultural development (EHEA, 2009). In particular, widening access to higher education is considered as a precondition of societal progress and economic development (EHEA, 2012).

The authors of Chapters 5, 6 and 7 present the situation in two EHEA countries: Austria and Lithuania. They discuss two different ways in which quality assurance has been used to stimulate transformation of HEIs towards ESD.

Although the Austrian and Lithuanian examples show different approaches to quality assurance and how they can be applied to initiatives in the area of sustainable development, they both explore the links between external (both international and national) and internal (institutional and/or study programme) dimensions. The ultimate goal of such quality exercises should, they say, be for universities themselves to become genuine 'learning organizations'.

In Austria, for example, internal and external quality management tools are considered as main drivers for change (Chapter 5, Shovakar & Bernhard). The coherence of external and internal steering criteria is formalized by the performance agreement between the Ministry for Science and Research and individual Austrian HEIs. Each performance agreement includes explicit initiatives of the EHEA (e.g. the Bologna Process) as well as for ESD. The Bologna process is seen as a top-down approach in Austria, while ESD is characterized mainly as a bottom-up one, expressed by a variety of region- and sector-specific projects related to different aspects of sustainability.

The case study from Lithuania (Chapter 6, Galkute) notes that common external evaluation areas and steering criteria are established, but that qualitative and quantitative indicators are specified by the internal quality management system of each institution according to its mission and strategic directions. Two spheres of action are identified in terms of the transformative potential of HEIs, in: (1) academic activities, and (2) outreach activities.

In Chapter 7 academics from the University of Graz (Zimmermann et al.) discuss a quality management system which involves activities in research, teaching, lifelong learning and administration. The emphasis is on incorporating quality enhancement in specific areas and establishing task-oriented tools for quality management which use the logic of the strategic planning cycle implemented as a continual process. A case study of the Regional Centre of Expertise (RCE Graz-Styria) is analysed as an example of applying forward-looking quality assurance focused on ESD and institutional sustainability.

In countries like Australia the government has established a national action plan for ESD and external quality audits are complemented by an extensive focus on locally operated IT-enabled, just-in-time, integrated internal quality assurance and improvement systems which draw on both benchmarked, time series quantitative and qualitative data (Chapter 11, Scott). In Canada the Canadian Institutional Research and Planning Association is moving in a similar direction and the Canadian Quality Network of Universities is fostering cross-institutional linkages and benchmarking, including in the area of ESD.

Emerging characteristics of quality assessment

To become truly transformative, quality assurance, assessment and improvement processes, as highlighted by Vettori & Rammel and

Galkute in Chapter 3 and 6, will need to help HEIs shift from being 'overly managerialistic and formalistic' and bureaucratic to become more development oriented and flexible by seeking data on areas for quality improvement and then by addressing them promptly and wisely. This, in turn, will require each HEI's leaders to work together to engage their staff and students with gathering quality tracking data, setting priorities, identifying solutions and assisting one another to test and enhance their selected improvement solutions under controlled conditions before scale-up (Chapter 4, Mader). This entails, therefore, the development of a *quality culture* where undertaking such quality tracking and improvement work becomes normal and is actively rewarded and supported by the HEI. The authors emphasize that quality has to be internalized and integrated into strategic management by going beyond stand-alone evaluation procedures. It is to be shared though multiple processes, through consistent stakeholder engagement and strong but shared leadership (Mader, 2012).

Placing transformational quality and its assurance as 'a core value of higher education institutions' instead of it being 'an externally imposed chore' (Vettori & Rammel) brings to the fore several critical challenges for HEIs interested in embracing this notion. One of these is how to design processes that not only assist in understanding of the values of the particular HEI but assessing them in a manner that contributes to the development of the organization in ways that are consistent with sustainability principles and which cover the ethical dimensions explicitly (Galkute, Chapter 6). The testament to the challenge is the fact that many assessments designed to 'capture' sustainability often fail to cover its ethical dimensions. In Chapter 9, Dahl shares the outcomes of EU projects that helped to develop methodologies and indicators for assessing educational initiatives like ESD. Making the values that underpin quality assessment and evaluation more explicit is seen as being a key step in enabling HEIs to become more change-capable and effective at consistently implementing and improving their innovations. This will, it is argued, lead to their conscious cultivation and, as a consequence, make educational processes more value-oriented.

Conceptualization of quality assessment as a transformative process also underscores the need for engaging with the multiple internal and external stakeholders concerned with moving universities and colleges to become more change- and ESD-focused. Brundiers et al. (Chapter 10), for example, stress the difficulties of maintaining constant engagement with key stakeholders in all stages of course development and in securing links between individual courses and educational programmes. This focus on transformation of every aspect of university operations again brings to attention the need for taking a whole institution approach and using a process of continuous assessment and improvement.

It is essential, say our authors, for HEIs to align their quality management systems to reflect their institution's vision and mission (Mader and Galkute in Chapters 4 and 6). This requires enabling stakeholders to become part of the quality management process in ways where they can most add value. When seen as a process of collective learning and appraisal, evaluation (making judgements of worth) can provide a unique opportunity to build and test an institution's vision for ESD and help ensure stakeholders become part of the process which can, in turn, encourage them to take responsibility for achieving the (shared and evolving) vision. Reflection, innovation and creating a shared vision with stakeholders can, the practitioners in this book argue, result in a qualitatively different, transformative, higher education quality process and set of outcomes.

This approach to quality assurance advocated in the chapters that make up the book acknowledges the multidimensional nature of transformation, the many value systems and interests involved, and the need to integrate the personal (Chapters 11 and 9 by Scott and Dahl), organizational (Chapter 3 by Vettori & Rammel) and societal (Chapter 9 and 12 by Dahl and Gapor et al.) dimensions of change.

It has already been noted that, in building a focus on continuous quality improvement and innovation through an evidence-based, networked and collegial process, HEIs will be modelling the exact process necessary to foster sustainable and resilient public and private instrumentalities and societies – that is, the values that have been found to underpin both change-capable individuals and change-capable institutions. Both acknowledge the need for constant innovation and adaptation in which communities of practice find common ground for engagement. Whilst the values that underpin ESD are seen by authors like Vettori and Rammel as 'guiding the quality goal itself', the development of a change-capable, sustainability-focused culture committed to continuous quality assessment and improvement is seen by all of our authors as being the bedrock for transforming our HEIs and the societies they serve for a sustainable future.

'Quality' as a central tool for the transformation of HEIs

As noted at the outset of this chapter, 'quality' is a term much used in many contexts, including in education at all levels across the world. A key objective of this book is to illuminate what is meant by this concept in the distinctive context of tertiary education for sustainable development and to provide proven and feasible ways in which to achieve it. It has been emphasized that, when evaluating quality in higher education, it is important to distinguish between quality as 'fitness for purpose' and quality as 'fitness of purpose'. The former concerns the importance of system alignment, the latter concerns making sure that we focus on what most counts for the sustainable future of our world.

The chapters that make up this book, when taken together, not only identify *what* needs to change in higher education in order to make it more suited to the sustainable development needs of the 21st century but also *how* this might be achieved in a way that is consistent, effective and continuously monitored and improved. Our contributors argue that the key role of our higher education institutions is not only to produce graduates for today but the sustainability-literate, change savvy, inventive, ethically robust leaders for tomorrow, and for this to happen a profound transformation of our HEIs is necessary.

It is, therefore, the effective management by our current universities of both the *what* and the *how* of transformation for sustainability in their curriculum, research, community engagement activities and ways of operating that will be most telling for the sustainable future we seek.

We have found that HEIs are increasingly using internal quality management systems as a lever for their systemic transformation towards becoming more change-capable, proactive, inventive and resilient, places in which 'quality' and continuous evaluation and collaborative improvement are not the sole province of a special unit but are the institutional norm, used by all to build a sustainable institution and, through this, able to contribute to a sustainable future. In such a shift quality control becomes quality research and an element of creative strategic management. This will require a profoundly different attitude to assessment that focuses not only on superimposed targets but is also helpful for improvement of higher education practices as a social learning to change process. Such evaluation will then embrace not only goals collectively defined and pursued by the stakeholders but also the process of collective learning in pursuing these goals.

It is in this way that, as we seek transformation towards sustainability in our HEIs, we not only advocate for a 'fitness for purpose' approach to quality but a 'fitness of purpose' approach in which the purpose is systematic, ongoing transformation towards sustainable development. We seek qualities, and quality systems, that actively develop *fitness for transformation*.

We trust that what is outlined in the book helps provide practical insights and guidance to help you address this critical higher education transformation agenda efficiently and effectively.

References

Barber M., Donnelly K., Rizvi S. (2013) *An Avalanche is Coming: Higher Education and the Revolution Ahead* (London: Institute for Public Policy Research).
Beck U., Cronin C. (2006) *The Cosmopolitan Vision* (Cambridge: Polity Press).
Dill D., Beerkens M. (eds) (2010) *Public Policy for Academic Quality: Analyses of Innovative Policy Instruments* (Dordrecht: Springer Publishers).
EHEA (2005) 'Standards and Guidelines for Quality Assurance in the European Higher Education Area', ENQA 4 March 2005.
EHEA (2009) 'The Bologna Process 2020 – The European Higher Education Area in the New Decade', Communiqué of the Conference of European Ministers Responsible for Higher Education, Leuven and Louvain-la-Neuve, 28–29 April 2009.
EHEA (2012) 'Making the Most of our Potential: Consolidating the European Higher Education Area', Bucharest Communiqué, Bucharest, 26–27 April 2012.
Florida R. (2002) *The Rise of the Creative Class: And How It's Transforming Work, Leisure, Community and Everyday Life* (New York: Perseus Book Group).
Fullan M., Scott G. (2009) *Turnaround Leadership for Higher Education* (San Francisco: Jossey Bass).
Mader C. (2012) 'How to Assess Transformative Performance towards Sustainable Development in Higher Education Institutions'. *Journal of Education for Sustainable Development*, 6 (1) 79–89.
Mader C., Scott G., Razak D. A. (2013) 'Effective Change Management, Governance and Policy for Sustainability Transformation in Higher Education'. *Sustainability Accounting, Management and Policy Journal*, 4 (3) 264–284.
Payyappallimana U., Fadeeva Z., O'Donoghue R. (2013) 'Traditional Knowledge and Biodiversity within Regional Centres of Expertise on Education for Sustainable Development'. In U. Payyappallimana, Z. Fadeeva (eds) *Innovation in Local and Global Learning Systems for Sustainability: Traditional Knowledge and Biodiversity – Learning Contributions of the Regional Centres of Expertise on Education for Sustainable Development* (Yokohama: UNU-IAS).
Rychen D. S. (2004) 'Key Competencies for All: An Overarching Conceptual Frame of Reference'. In UNESCO (ed.) *Developing Key Competencies in Education* (UNESCO: International Bureau of Education), 5–34.
Scott G., Coates H., Anderson M. (2008) *Learning Leaders in Times of Change: Academic Leadership Capavilities for Australian Higher Education* (Sydney: University of Sydney and Australian Council for Educational Research).
Scott G., Tilbury D., Sharp L., Deane E. (2012) *Turnaround Leadership for Sustainability in Higher Education* (Sydney: Office of Learning and Teaching, Australian Government).
Sen A. (1999) *Development As Freedom* (New York: Knopf).
Spohn W. C. (2003) *Reasoning from Practice*. Syllabus Narrative for the Carnegie 'A life of the mind for practice' seminar (Stanford, CA: The Carnegie Foundation for the Advancement of Teaching).

Tiana A. (2004) 'Developing Key Competencies in Education Systems: Some Lessons from International Studies and National Experiences'. In UNESCO (ed.) *Developing Key Competencies in Education* (UNESCO: International Bureau of Education), 35–80.

UN Economic Commission for Europe (UNECE) (2011) 'Learning for the Future: Competences in Education for Sustainable Development', ECE/CEP/AC.13/2011, April 2011, http://www.unece.org/fileadmin/DAM/env/esd/ESD_Publications/Competences_Publication.pdf, date accessed 15 July 2014.

UNESCO-CEPES (2007) 'Quality Assurance and Accreditation: A Glossary of Basic Terms and Definitions', http://unesdoc.unesco.org/images/0013/001346/134621e.pdf, date accessed 2 May 2014.

United Nations (1992) *Agenda 21: Programme of Action for Sustainable Development* (New York: UN Department of Public Information).

Part I
Transformation of Higher Education in Changing Society: Implications for Quality Management

2
Rankings and the Reconstruction of Knowledge during the Age of Austerity

Ellen Hazelkorn

A new world order?

The severity of the ongoing global economic crisis and the continuing shift to the knowledge-based economy have put considerable pressure on higher education to play a greater role in national economic recovery, and to demonstrate greater relevance to, and better value for, individuals and society. Despite commitments to/for institutional autonomy, there is increasing evidence of greater government steering and regulation of the higher education and research system. In some countries, there is growing emphasis on higher education as an arm of industrial/economic policy rather than its attributes for human capital development. Some governments are able to invest heavily or at least retain their level of investment (often as part of stimulus) in higher education and R&D while others face serious financial strain. The rising influence of global rankings has drawn the world's attention to disparities, while at the same time making the higher education world immediately more competitive and multi-polar. As a result, there is evidence of a widening gap in 'world-classness' between nations and institutions. There are several aspects to this dynamic.

The application of knowledge is widely acknowledged as being the source of social, economic and political power. Studies repeatedly show the strong correlation between educational attainment and social and economic advantages for individuals and society (OECD, 2009). Because its contribution to economic growth can be so significant, higher education has been called 'the engine of development in the new world economy' (Castells, 1994, p. 14). This link is especially acute as countries respond to the global economic crisis, with international organizations, such as the OECD (2010) and World Bank (2009), urging governments to protect education and research budgets as the key to economic recovery. Yet, even before the current crisis, globalization had been forcing change across all knowledge-intensive industries creating a 'single world market', and sharing many characteristics of other multinationals. Because knowledge production transcends national boundaries, university[1] membership of global networks has become an essential device for extending institutional reach, accessing a wider pool of talent and funding opportunities, and providing outlets for research products and services (Maslen, 2013). In the global economy, national pre-eminence is no longer sufficient.

The severity of the 'fiscal crisis of the state' (O'Connor, 1973) has cast a critical light on the alarming costs of higher education, and the often impenetrable world of university and academic language. No doubt it suited higher education to affirm its key position vis-à-vis economic prosperity because the claim underpinned substantial hikes in public expenditure throughout the 'golden years' preceding the global financial crisis (GFC) of 2008. Yet, if higher education is the engine of the economy, governments are now asking for verifiable evidence – and they are not the only ones. While primary and secondary education is usually referred to as compulsory education, higher education is seen to encompass both public and private benefit of which the latter is increasingly to the fore. By subjecting higher education to the market, students are acting as other consumers, questioning the value-for-money as set against the tuition fee paid or institutional status/reputation as defined by rankings. Evidence of quality and the pursuit of excellence are the key mantras now dominating higher education, inside and outside the academy.

Another trigger has been the emergence and rising prominence and obsession with global rankings. Rankings have linked the investment attractiveness of nations with the talent-catching and knowledge-producing capacity of higher education. Recognizing this significance, the Irish Minister for Education and Science, speaking on behalf of the European Council, said the 'news is not all that good' that too few European universities featured among the world's top 500 (Dempsey, 2004) – especially for a region which sought to be the 'the world's most competitive and dynamic economy by 2010' (Europa, 2000). Others argued Europe's universities stood at a crossroads: 'the recent publication of global rankings...has made most policymakers aware of the magnitude of the problem and sparked a public debate on university reform' (Dewatripont, 2008, p. 6; Hazelkorn and Ryan, 2013).

This response is not restricted to Europe (Hazelkorn, 2013a). Today, few countries have not spearheaded a review of their higher education system; fewer still have refrained from identifying and then resourcing a few universities as world-class. An emerging global model (EGM) (Mohrman et al., 2008) of higher education is being developed by governments and promoted by universities and academics who seek to adopt or ape the characteristics of the top 20, 50 or 100 universities, using the criteria promulgated by global rankings. The world-class research university has become the panacea for ensuring success in the global economy and world science while reinforcing or imposing steeper hierarchical differentiation between selective elite research universities and recruiting mass teaching institutions. Narrowing the policy framework in this way has not just affected debate at the system level. Because the language of rankings has insinuated itself into public discourse at almost every level of decision-making and public commentary, it has also influenced our understanding of knowledge, how new ideas are generated, and by whom, and how we access it.

This chapter will explore these issues examining the way in which rankings are helping to reconstruct a hierarchy of knowledge and disciplinary values which privileges a particular world order and type of institution. And, because rankings give the appearance of scientific rigour, they have supported the trend towards the quantification of performance as an indicator of quality – in other words, policy-making by numbers. The first section will explain how rankings measure knowledge, while the following section will contrast that configuration with contemporary understanding of knowledge production; the third section will look at the impact this is having on the academy while the fourth section will look at the influence on policy. Finally, the conclusion will address the wider implications for higher education and knowledge production.

How rankings measure knowledge

Rankings are the latest form of accountability and transparency instrument for measuring higher education performance; other mechanisms include, inter alia, college guide books, accreditation procedures either by government or professional organizations, institutional assessment and evaluation processes, benchmarking and classification systems (Hazelkorn, 2011). Global rankings have become an international phenomenon since 2003; there are four main phases. According to Webster (1986, p. 14, pp. 107–119), the man who invented rankings was

James McKeen Cattell; in 1910, he published *American Men of Science* to show the '"scientific strength" of leading universities using the research reputation of their faculty members'. *US News and World Report* Best College Rankings (USNWR) marks the second phase in 1983, providing consumer-oriented information to students and parents. Its rise to prominence coincided with the ideological and public 'shift in the Zeitgeist towards the glorification of markets' (Karabel, 2005, p. 514). Over the decades, the range of national rankings has continued to expand, and the number is growing (Salmi and Saroyan, 2007; Usher and Jarvey, 2010; Hazelkorn, 2012a).

The third era was marked by publication of the Shanghai Jiao Tong Academic Ranking of World Universities (ARWU) global ranking in 2003, quickly followed by, inter alia, Webometrics (produced by the Spanish National Research Council), Times Higher Education-QS World University Ranking (THE-QS) 2004–2009, Taiwan Performance Ranking of Scientific Papers for Research Universities (HEEACT) in 2007 (now called the National Taiwan University Ranking (NTU)), The Leiden Ranking (2008) by the Centre for Science and Technology Studies (CWTS) and SCImago (2009) by a team of Spanish researchers. More recently, the QS World University Rankings (2010) and the Times Higher Education World University Ranking (THE) (2010) have emerged as individual rankings, the latter in partnership with Thomson Reuters, thus representing a significant intervention into the market by the producer of one of the major world bibliometric databases. The EU has also entered the global intelligence information business, launching U-Multirank[2] in May 2014 as a sister instrument to its classification tool, U-Map.

The fourth phase is marked by the intervention of supra-national authorities such as the above-mentioned EU, and the OECD with its *AHELO* project which sought to assess higher education learning outcomes. Alongside the US federal government's forthcoming Postsecondary Institution Rating System (PIRS), these developments mark a very significant paradigm shift. As globalization and marketization of higher education accelerates, and international mobility increases, governments have been compelled to step-in to regulate the marketplace. Concern is not simply about education per se but about the security of the global economy. As of writing, there are nine active global rankings, and over 150 other global, national and discipline/specialist rankings, experiencing varying degrees of popularity, reliability and trustworthiness (see Box 2.1; Hazelkorn, 2015).

> *Box 2.1* Major global rankings, 2014 (in order of date of origin)
>
> - Academic Ranking of World Universities (ARWU) (Shanghai Jiao Tong University, China), 2003
> - Webometrics (Spanish National Research Council, Spain), 2004
> - National Taiwan University Rankings (formerly Performance Ranking of Scientific Papers for Research Universities, HEEACT), 2007
> - Leiden Ranking (Centre for Science & Technology Studies, University of Leiden), 2008
> - SCImago Journal and Country Rank (SJR) (Spain), 2009
> - University Ranking by Academic Performance (URAP) (Informatics Institute of Middle East Technical University, Turkey), 2009
> - QS World University Rankings (Quacquarelli Symonds, UK), 2010
> - *Times Higher Education* World University Ranking (*Times Higher Education*, UK), 2010
> - U-Multirank (European Commission, Brussels), 2014
>
> *Note*: This list only includes active global rankings.

Rankings compare higher education institutions according to a range of indicators weighted in line with criteria determined by the ranking organization; there is no such thing as an objective ranking. Scores are aggregated into a single digit to identify the best institution. Rankings are essentially one-dimensional since each indicator is considered independently from the others, whereas in reality 'multicollinearity is pervasive' (Webster, 2001, p. 236); in other words, older well-endowed private universities are more likely to have better faculty/student ratios and per student expenditure compared with newer public institutions. Data is drawn from a combination of sources, including, for example, government databases, university sources, and survey data of students, peers, employers or other stakeholders (Usher and Medow, 2009, p. 6). There are advantages and disadvantages to each data form and indicator, but for the purposes of this chapter and argument, the focus will be on measuring knowledge and universities as the primary vehicle for knowledge production (for extensive discussion see Hazelkorn, 2011 and forthcoming 2015, chapter 2).

Global rankings depend on internationally comparative data but this can be complex and imperfect; national contexts resist attempts to make simple and easy comparisons and there is a serious lack of consistency in data definition, collection practices and reporting – even within national borders. Arguably this has led global rankings to focus disproportionately on research because of availability of such data. On the other hand, global rankings use research and scholarly productivity as a proxy for academic quality. When research and research-dependent indicators are combined (e.g. academic reputation, doctoral students and awards, research income, citations, academic papers, faculty and alumni medals and awards, internationalization), the research component rises: for example: NTU (100%), URAP (100%), ARWU (100%), THE (93.5%) and QS (70%). Taking account of the multicolinearity between variables, this illustrates the extent to which rankings are essentially a measure of research, and all the other indicators simply cloud this fact.

Rankings, similar to other research assessment processes, rely heavily on the bibliometric data supplied by Thompson Reuters' Web of Science (WoS) and Elsevier's Scopus – and are increasingly seen as problematic as the following discussion illustrates. They primarily collect peer-reviewed articles, from approximately 12,000 and 22,000 journals respectively; despite expansion to open access journals and conference proceedings, the databases capture only a small proportion of research activity. These methods effectively privilege research undertaken by the physical, life and medical sciences as more important or valuable than the arts, humanities and social science (AHSS) journals. Focus on publications in high-impact journals, as determined by the average number of citations per article (Garfield, 1955 and 2005), also assumes journal quality is equivalent to article quality. ARWU awards 20 per cent of its score to just two publications, *Science* and *Nature*, while SCImago uses the journals' scientific prestige. Many governments and research agencies have adopted this practice; the Australian Research Council (ARC, 2010) has developed a controversial four-tier structure whereas the European Science Foundation ERIH project (European Reference Index for the Humanities) classifies journals in 15 fields according to reputation and international reach (ESF, 2010).

While the 'hard' sciences publish frequently in journals with multiple authors, AHSS disciplines publish or disseminate findings in a variety of formats, such as books, book chapters and conference proceedings, policy reports, contribution to technical standards, electronic formats or open source publications, prototypes and artefacts, musical compositions, or other forms of knowledge transfer/exchange. International databases favour English-language publications because it is considered the universal language of science, a practice viewed as disadvantaging the social sciences and humanities because they regularly consider issues of national or local relevance and publish in the national language. Such practices can also disadvantage the natural and physical sciences as they seek to solve national-oriented problems of disease, aquaculture, agriculture, and so on. New research fields, interdisciplinary research that crosses traditional academic boundaries or ideas that challenge orthodoxy can also find it difficult to be published, while (in contrast) papers containing arguments with significant errors may regularly be cited because they are illustrations of bad practice. Similarly, measuring research income benefits capital-intensive bio-sciences and medicine disciplines, but says little about the impact of research on teaching or more broadly on society and the economy. Indeed, rewarding input factors, such as income earned, without correlating for outputs, ignores issues about efficiency, for example, which would be a shared concern of both governments and institutions nowadays. Another difficulty is that traditional methodologies measure impact solely in terms of citation counts, which means they are unable to capture the way in which higher education and research makes an impact beyond the academy, for example on regional or civic engagement or on social, cultural, economic and environmental sustainability – both major policy objectives for many governments and HEIs.

To get around some of these problems, WoS has joined with Google Scholar to capture grey literature (GreyNet) and with SciELO and RedAlyC to capture Latin American authors; Scopus is also developing techniques to capture AHSS research. Initiatives such as Altmetrics, in addition to grey literature, facilitate mentions in social and news media, references in databases and repositories, downloads, and so on (Alperin, 2013). There is also a growing movement towards open access and open science, driven by reaction against the level of control exercised by publishers and a broader policy perspective that the results of publicly funded research should be made widely available to enhance public knowledge but also spur knowledge exchange and innovation (Dunleavy, 2013). Thus, web-based interfaces, such as Google Scholar, institutional repositories or other web-based technologies widen also begin to challenge the notion that impact and benefit is measured only by readership among scholarly-scientific peers rather than on society and the economy more broadly.

Understanding knowledge and knowledge production today

The organization of higher education has traditionally reflected a simplistic understanding of knowledge creation, social class, and skill/labour market requirements. Over time, these boundaries have blurred as labour markets have matured, professional/academic disciplines have moved up the value chain, and knowledge production has moved beyond the academy. In contrast to the classical university of Humboldt and Newman whose mission distinguished teaching and research from commercial activity, US Land Grant universities of the late 19th century focused on teaching agriculture, science and engineering as a response to the industrial revolution on the basis that knowledge was sought for the public good. About the same time, in the UK, '[l]ocal entrepreneurs and civic leaders responded to the needs for scientific knowledge and a healthy and skilled workforce by founding universities to underpin the economic success of the cities in the nation's heartland' (Goddard, 2009, p. 6). In the late 20th century, the birth of a new generation of universities and other higher education institutions catering for a wider range of socio-economic and learner groups, educational requirements, and new careers and professions marks the birth of the phase of universal access in response to changes in society and the economy. Today, institutions established in the post-1970s represent the majority of all HEIs in the world.

Accordingly, the role of higher education has changed dramatically, from institutions attended by a small cohort of intellectual or social elite to one where attendance has become more or less obligatory for a wide range of occupations and social classes. As knowledge is recognizably more complex and the knowledge economy more demanding, many characteristics of the newer institutions, for example closeness to industry and the professions, application-focused and socially responsible research and commercialization, and engagement with the wider community, have been promoted by government and adopted by all universities, regardless of origin. At the same time, external pressure to diversify funding through 'third stream' activities has also pushed universities to reconfigure their organizations and develop 'boundary crossing' mechanisms and units which ensure active interaction with external stakeholders, the community and region (Hazelkorn, 2010; Ward and Hazelkorn, 2012). Today, institutional differentiation is increasingly multidimensional; distinctions according to mission and core tasks, for example teaching and research, are interwoven with diversity among the student and faculty cohort or programme and pedagogical profiling (Reichert, 2009). Goddard's (2009) concept of the 'civic university' talks of the research university making a multifaceted contribution to the economy, as a source of knowledge and skilled employees, and as the centre for regional economic clusters, leveraging the advantages of place; the EU 'knowledge triangle' links the three elements of education, research and innovation (Europa, 2010). These developments call into question the logic of terminology such as 'mission drift', and usage of the legal title 'university' as a differentiator in the higher education sector (Dill and van Vught, 2010, p. 3). Excellence within a field(s) of specialization is becoming a more significant indicator of a university's position rather than research intensity; or as Terenzini et al. (2010, p. 22) argue, what colleges and universities 'do' is more important than what they are 'called' (Hazelkorn, 2012b).

Traditionally, research activity – and commensurately types of higher education institutions – has been divided into a hierarchy of functions whereby basic research was considered the primary and most important form of scholarship. Since 1963, the OECD *Frascati Manual* – version 7.0 is currently underway – (OECD, 2002) has provided a basic definition of research, distinguishing between basic, applied and experimental production. Challenging this narrow frame, Boyer (1990) highlighted four different 'scholarships': discovery, integration, application and teaching (cf. Delanty, 2001, p. 9). He argued that the word 'research' is a relatively recent concept having displaced the broader term 'scholarship': 'What we now have is a more restricted view of scholarship, one that limits it to a hierarchy of functions. Basic research has come to be viewed as the first and most essential form of scholarly activity, with other functions flowing from it' (Boyer, 1990, p. 15).

Revolutionary as his argument was at the time, Boyer's conceptualization still centred on a linear knowledge-push understanding, in other words that knowledge and expertise was produced within the academy with a uni-directional flow of this expertise *out* of the academy (O'Meara and Rice, 2005).

Gibbons et al. (1994) went further, arguing that as knowledge became more complex it gave rise to new disciplines, methodologies and ways of thinking. Unlike traditional Mode 1 research, which is disciplinary or 'curiosity-oriented', conducted by individuals in a secluded/semi-secluded environment and achieving accountability through peer-review, Mode 2 knowledge is 'socially robust', collaborative and interdisciplinary. It is focused on useful application, with external partners including the wider community, and achieves accountability and quality control via social accountability and reflexivity. Mode 2 moves the site of problem formation, investigation, discovery and resolution into the public realm or agora: '[The] agora is the space in which societal and scientific problems are framed and defined, and where "solutions" are negotiated. It is the space, par excellence, for the production of socially robust knowledge' (Gibbons, 2002, p. 59).

> *Box 2.2* Changes in our understanding of knowledge production
>
> - Research aligned with social and economic progress (Vannevar Bush, *Science The Endless Frontier*, 1945), focused on fundamental scientific research, excluding the humanities and social sciences.
> - Basic, applied and experimental production (OECD, Frascati Manual, 1963, 1st edn; 2002, 6th edn, p. 28): 'comprises creative work undertaken on a systematic basis in order to increase the stock of knowledge, including knowledge of man, culture and society and the use of this stock of knowledge to devise new applications'.
> - *Engaged scholarship* (Ernest Lynton, 1987; Ernest Boyer, 1990, 1996): 'Knowledge generation is a process of co-creation, breaking down the distinctions between knowledge producers and knowledge consumers', defining the research problem, choosing theoretical and methodological approaches, conducting the research, developing the final products, and participating in peer evaluation. (Saltmash et al., 2009, p. 10)
> - *New production of knowledge* (Michael Gibbons et al., 1994): Mode 2 is 'socially robust' and interdisciplinary knowledge, created within the context of being useful for the resolution of specific problems, in contrast to traditional knowledge production (Mode 1) which is disciplinary or 'curiosity-oriented', usually conducted by individuals in secluded/semi-secluded environments.
> - *Grand challenges* (Graham, US Office of Science and Technology, 1987; National Academy of Sciences, 2004): problems of economic and social importance which are demonstrably hard to solve, thereby requiring improvements of several orders of magnitude, collaborative solutions and interlocking knowledge and innovation systems.

As the interconnectedness between higher education and society deepens, research 'comes increasingly to the attention of larger numbers of people, both in government and in the general public, who have other, often quite legitimate, ideas about where public funds should be spent, and, if given to higher education, how they should be spent' (Trow, 1974, p. 91; see also Lynton, 1994). This is the situation today, where governments and their publics are asking many questions about expenditure but also about value, impact and benefit; this has led to Mode 3 (see Hazelkorn, 2012b, p. 843) knowledge production which is required to achieve both social and public accountability.

This understanding has encouraged the problematization of research questions as 'grand challenges', such as climate change, human health and healthy living, food, energy and water security or sustainable cities. Owing to their scale and complexity, major social and economic problems transcend borders and disciplines, necessitating new methodological and organizational frameworks. They require collaborative solutions

Table 2.1 Knowledge production accountability

Mode 1: Peer accountability	Mode 2: Peer and social accountability	Mode 3: Social and public accountability
Pursuit of understanding of fundamental principles focused around 'pure disciplines' and arising from curiosity, with no (direct or immediate) commercial benefits. Conducted by a limited number of research actors in a secluded/semi-secluded environment. Achieves accountability via peer-review process. (Gibbons et al., 1994)	Pursuit of understanding of principles in order to solve practical problems of the modern world, in addition to acquiring knowledge for knowledge's sake. Broad range of research actors across breadth of disciplines/fields of enquiry. Achieves accountability via a mix of peer and social accountability. (Gibbons et al., 1994)	Research is focused on solving complex problems via bilateral, inter-regional and global networks, not bound by borders or discipline. Knowledge production is democratized with research actors extending/involving 'beyond the academy'. Emphasis is on 'reflective knowledge' co-produced with and responsive to wider society, with an emphasis on impact and benefit. Achieves accountability via social and public accountability. (Hazelkorn, 2012b, p. 843)

Source: Hazelkorn, 2012b.

and interlocking innovation systems, underpinned by interdisciplinary research teams working inter-institutionally and globally.

Interdisciplinary thinking is rapidly becoming an integral feature of research as a result of four powerful 'drivers': the inherent complexity of nature and society, the desire to explore problems and questions that are not confined to a single discipline, the need to solve societal problems, and the power of new technologies. (National Academy of Sciences, 2004, p. 2)

These developments illustrate how conceptions of knowledge production have leveraged ideas about clusters (Porter, 1985), innovation systems (Lundvall, 1992; Nelson, 1993), 'triple helix' (Etzkowitz and Leydesdorff, 1997) and 'quadruple helix' (Carayannis and Campbell 2009) to help reconfigure the role of the university within society. Dill and van Vught (2010, p. 9) argue that the broader national innovation system (NIS) 'perspective emphasizes the interactive character of the generation of idea, scientific research, and the development and introduction of new products and processes'.

These arguments challenge the linear science-push framework popularized in the post-WW2 era, and (attempt to) end fragmentation of the knowledge system. Knowledge production is a continuum or end-to-end process involving the whole process of discovery, spanning the spectrum from curiosity-driven to use-inspired, from blue-sky to practice-based; rather than basic versus applied knowledge, there is 'applied and not yet applied' (Boulton and Lucas, 2008, p. 9; see also Rothwell, 1994). Translational research, traditionally associated only with the bioscience's concept 'from bench-to-bedside', is now broadly associated with other disciplines seeking, for example, to close what the World Health Organization (2009, pp. 12–13) calls the 'science–policy gap'. By enveloping the whole innovation chain, higher education becomes a critical component of the eco-system, no longer simply the 'engine'. This transformation has helped democratize knowledge production, challenging the privilege of the 'ivory tower'; the end-user is both an active participant shaping the research agenda, and an assessor of its value, impact and benefit. In other words, knowledge has 'ceased to be something standing outside society, a goal to be pursued by a community of scholars dedicated to the truth, but is shaped by many social actors under the conditions of the essential contestability of truth' (Delanty, 2001, p. 105).

This conceptualization of knowledge stands in stark contrast to the vision promulgated by rankings, and used by both the academy and policy-makers to assess and reward research and academic performance and productivity. While U-Multirank seeks to capture a wider range of indicators about higher education performance, it still maintains a conventional understanding of research and its outputs. It is particularly ironic that at a time in which governments, including the EU (Europa, 2013), require university-based research to demonstrate impact and relevance, they should use methodologies that continue to reinforce a traditional approach to knowledge production and dissemination (Hazelkorn, 2009).

Rankings and the academy

Rankings have had a profound impact and influence on higher education and the policy community despite protestations to the contrary (Hazelkorn, 2011). Users of rankings extend beyond the original target group of students and their parents, and now include, inter alia, policy-makers, employers, foundations and benefactors, potential collaborators and partners, alumni, other HEIs and many other stakeholders. Public opinion has been greatly informed by rankings' simple and simplistic message, and the language of rankings has insinuated itself into public discourse at almost every level of decision-making and public commentary around the world.

However, higher education institutions and individual academics are not innocent victims in this process. Evidence from around the world shows how rankings have had a significant impact and influence on the business of higher education, including the choice of academic partners (Hazelkorn, 2011). While some HEIs deliberately strive to improve their standing in the rankings, other simply wish to be included – because being ranked is equivalent to being visible to potential students, HE partners, policy-makers, the media, and so on. This explains why HEIs advertise on the web-pages of the various rankings, and use Facebook, Twitter and other forms of social networking. As a result, universities 'at all levels along the selectivity spectrum' have sought to cushion themselves by introducing subtle and major changes to their mission, organization and priorities to boost/protect their position and reputation and/or influence their rank (Ehrenberg, 2001, p. 146). Senior higher education 'administrators consider rankings when they define goals, assess progress, evaluate peers, admit students, recruit faculty, distribute scholarships, conduct placement surveys, adopt new programs and create budget' (Espeland and Sauder, 2007, p. 11).

Table 2.2 below links particular actions taken by HEIs with the weightings attributed to specific indicators by different rankings. Thus, institutional actions might typically include institutional restructuring and the reorganization of research through, for example, the creation of research institutes and graduate schools, and increasing research output and publications in international journals, etc. with the objective of improving the citation index. Because bibliometric methodologies favour the bio-sciences, these disciplines often receive preferential treatment, targeting indicators, for example, research output, research quality/citation index. Many universities aggressively headhunt international scholars, targeting indicators, for example, research output, quality and proportion of international faculty, and generously reward those who publish in highly cited journals, targeting indicators, for example, research output. Others have measures to positively affect their faculty–student ratio, targeting indicators, for example, teaching quality. Many have decisively improved their marketing and public communications using expensive and extensive advertisement features, for example in *Nature* or *Science*, glossy brochures or marketing tours in an effort to influence peer appraisals or improve recruitment.

Table 2.2 Indicative mapping of institutional actions against selective rankings (2014)

	Examples of actions taken by HEIs in response to rankings	Approximate weighting
Research	• Increase output, quality and citations • Recruit and Reward faculty for publications in highly cited journals • Publish in English-language journals • Set individual targets for faculty and departments • Increase number/proportion of PhD students	• ARWU = 100% • THE-QS = 60% • NTU = 100% • THE = 93.5% • QS = 70%
Organization	• Merge with another institution, or bring together discipline complementary departments • Incorporate autonomous institutes into host HEI • Establish centres of excellence & graduate schools • Develop/expand English-language facilities, international student facilities, laboratories, dormitories • Establish institutional research capability • Embed rankings indicators as a performance indicator or as a contract between the presidency and departments • Form task group to review and report on rankings	• ARWU = 10%; • Research related indicators as above
Curriculum	• Harmonize with EU/US models • Favour science/bio-science disciplines • Discontinue programmes/activities which negatively affect performance • Grow postgraduate activity relative to undergraduate • Withdraw from programmes which do not enhance research intensity • Positively affect faculty-student ratio • Improve teaching quality • Incorporate results in development of new study programmes and degrees	• THE-QS = 20% • THE = 30% • QS = 30%

Students	• Target recruitment of high-achieving students, esp. PhD students • Offer attractive merit scholarships and other benefits • More international activities and exchange programmes • Open international office and professionalize recruitment	• THE = 9.25% • QS = 5%
Faculty	• Recruit/headhunt international high-achieving/HiCi scholars • Create new contract/tenure arrangements • Set market-based or performance/merit-based salaries • Reward high achievers • Identify weak performers • Enable best researchers to concentrate on research/relieve them of teaching	• ARWU = 80% • THE-QS = 95% • NTU = 100% • THE = 97.5% • QS = 95%
Public image/ Marketing	• Reputational factors • Professionalize admissions, marketing and public relations • Ensure common brand used on all publications • Advertisements in *Nature* and *Science* and other high-focus journals, • Highlight medals won by staff/alumni • Expand internationalization alliances and membership of global networks	• ARWU = 50% • THE-QS = 40% • QS = 50% • THE = 33%

Source: Hazelkorn, 2011, pp. 192–193 [updated].

Universities everywhere are preoccupied with recruiting high-achieving students, preferably at PhD level, who, like international scholars, will be assets in the reputation race, targeting indicators, for example quality of faculty, proportion international, research output, research quality/citation index, peer appraisal, graduate employability. While many of these actions are inevitable strategic decisions in a competitive world, there is a suspicion that 'rankings are always in the back of everybody's head' (Espeland and Sauder, 2007, p. 11).

Implications for higher education policy

Rankings were initially developed as a consumer product but have quickly become a management tool and policy instrument. In the 'age of austerity' rankings have coincided with the acceleration of competition and reinforced the trend for quantification as affirmation of quality and value-for-money. These developments are having a profound impact on higher education. They privilege world-class research-intensive universities, and traditional forms of knowledge creation. Bizarrely, the emphasis is on activities which lead to high rank despite the fact that the indicators neither reflect contemporary understandings of the spectrum of knowledge production nor measure what public policy seeks to achieve. This juxtaposition highlights deepening policy tensions and contradictions.

Policy-makers reacted swiftly to what they perceived global rankings to be saying about the quality and performance of their higher education systems and, accordingly, about their economies. As concerns about global competitiveness and budgetary constraints have risen, many governments have become less concerned with supporting educational diversity and more with excellence. This is reflected in the way in which the language of rankings and the actual indicators have been incorporated into policy discourse and processes. Rankings have also sharpened national debate about assessment and the measurement of performance and productivity. While higher education has always been competitive, performance against international benchmarks is now regularly linked to resource allocation, and has underpinned accountability or transparency strategies and instruments aiming to measure return on investment and value-for-money. In many instances governments have directly adopted or 'folded-in' the indicators used by rankings into their own processes or used rankings to set targets for system restructuring.

For many governments, the world-class university has become the key for success, acting as a magnet for mobile investment and talent. Oblivious to the huge resources gap between different higher education systems, and between public and private universities, many governments are using rankings (explicitly and implicitly) to drive change in an effort to achieve world-class status. Because size matters, governments have sought to restructure their higher education and research systems and selectively fund excellence. France, Germany, Denmark, Russia, Spain, China, South Korea, Taiwan, Malaysia, Finland, India, Japan, Singapore, Vietnam and Latvia – among many other countries – have all launched initiatives with the primary objective of creating world-class universities, using the ARWU or the THE-QS definition (Hazelkorn, 2011; Mohrman et al, 2008). US states have similarly sought to build or boost flagship universities, elevating them to what is known as Tier One status, a reference to *US News and World Report* college rankings (Arnone, 2003; Lederman, 2005; Ludwig and Scharrer, 2009). Ironically, the attention now being given to world-class or flagship universities is occurring at the same time that student and labour-market demand for graduates is escalating (Hazelkorn, 2013b). Can concentration of resources and talent in a handful of elite resource and research-intensive universities deliver the required results or will it ultimately undermine capacity across all society?

Policy tensions are also evident in the way in which knowledge production is understood and measured. Over the years and especially in response to the prolonged economic crisis, higher education has been asked to respond more directly to societal needs, and to engage in sustained, embedded and reciprocal engagement beyond campus walls, discovery that is useful beyond the academic community and service that directly benefits the public. Different organizational models and initiatives are emerging which bring together actors from civil society, the state and state agencies to mobilize and harness knowledge, talent and investment to address a diverse range of problems and need through coordinated action; there is increasing focus on technology and knowledge transfer, science parks, collaborative projects, work-based doctoral studies, and so on. Even researchers admit that:

> Researchers who benefit from opportunities in cities should ask what they can give back. More than half of the world's people live in cities, and that number is growing rapidly. So if scientists want to help the majority of the population, they need to turn their attention to urban areas. (Editorial, 2010)

However, once research is seen to have value and impact beyond the academy, there are implications for the organization and management of research at the national and institutional level, what kind of research is funded, and how it is measured and by whom. This has exposed a tension between research as vital for human capital development versus its contribution to economic development; between an emphasis on researcher curiosity versus alignment with national priorities; between funding excellence wherever it exists versus targeting funding to strengthen capability or build scale; between encouraging new and emerging fields and higher education institutions versus prioritizing existing strengths; and between valuing and funding traditional inputs and outputs (e.g. quantum of resources, scientific-scholarly articles and citations) versus impact and benefits (e.g. the contribution of research outcomes for society, culture, the environment and/or the economy).

Ironically, at a time when society requires higher education to meet a wider range of needs, rankings concentrate on only one end of the research spectrum as the only plausible measure of knowledge. In so doing, they misrepresent and pervert the research/innovation process by reinforcing a simplistic science-push view of innovation. Because the fundamental end of the spectrum is dominated by the bio-sciences, rankings effectively ignore the contribution, for example, of the creative/cultural industries to innovation or the way in which social innovation brings about fundamental change to the social economy via new forms of mutual action, new ways in which economies can be managed, new forms of consumption, and the organization and financing of government. They fetishize a narrow definition of research, and fail to give adequate recognition to the breadth of higher education's contribution to society and the economy. Despite pronouncements about the 'knowledge triangle', 'third mission' or engagement, by over-emphasizing research, rankings narrowly define higher education.

Conclusion

At a time when society is asking more and more questions about the value of public investment in higher education and research, global rankings proselytize a single form of excellence based upon the US (not-for-profit) private research university experience (Tierney, 2009). They propel us backwards in time, reinforcing elite notions of a knowledge hierarchy with geo-political implications. Unconsciously, many of the policy decisions being taken rely heavily on normative assumptions of higher education, and research quality and performance. There are four interrelated tensions that emerge from this discussion:

1. Hierarchically ordering or stratifying knowledge – for example, through the practice of ranking journals – reinforces the status of elite institutions and a handful of countries as the primary knowledge producers

and generators of intellectual property. As knowledge production is fetishized in this manner, it is perceived as something that only some universities and academics can undertake and interpret.
2. The quantification of performance gives the 'appearance of scientific objectivity' (Ehrenberg, 2002, p. 147) whilst creating a powerful set of ideological and hegemonic ideas or values around which a particular model of higher education or concept of quality or excellence becomes the accepted norm. The debate as to which ranking is better should be seen as part of a global power struggle about who (which countries and institutions) should participate in world science.
3. Given serious constraints on government budgets coupled with accelerating global competition, there is increasing pressure on governments to target funding. In the same way that individual HEIs have identified those indicators most likely to improve global standing, many governments have chosen to prioritize science and technology research with converse results for the arts, humanities and social sciences. Because university-based research operates within the educational eco-system, imbalances in resources have broader implications for higher education.
4. Despite policy drivers that seek to place higher education and knowledge production within a dynamic societal eco-system, rankings emphasize individual institutional performance at the global level. This places a particularly high premium on the reputation and status of flagship research-intensive universities rather than the capacity of the higher education system or knowledge society. It encourages universities to bypass their country/region as they seek global recognition. This has implications not just for how HEIs relate to their region and but also how they relate and work with each other.

As competition for limited 'positional goods' intensifies and the world becomes increasingly multi-polar, governments struggle for their share of the global economic pie. Governments and institutions have sought refuge in rankings as a global indicator of quality without considering the implications of the policy trade-offs being made as a result (Hazelkorn, 2013b). No doubt, identifying better ways to value the full impact or benefit of higher education is difficult, but the urgency lies, to paraphrase Einstein, in finding ways to measure what counts rather than what is easy. Strangely, this presents higher education with an opportunity to help frame the new requirements in ways which do not undermine the values of higher education. It's time to push out the debate.

Notes

1. This refers to all doctoral-awarding higher education institutions (HEIs) irrespective of name or status in national law.
2. In contrast to existing rankings, U-Multirank is based on the principles of being: (i) user-driven, (ii) multidimensional, (iii) peer-group comparable, and (iv) multi-level (van Vught and Ziegele, 2011 and 2012). In this way, it promotes a broader range of quality measures than those used by existing rankings, and a 'build-your-own' system rather than an overall ranking for each institution or a single ranking of all the 18,000 HEIs in the world. Criticism has focused on the time-consuming nature of the process, the sunburst imagery used to display results, and appropriateness of the actual name (Hazelkorn, 2013c). Another EU project aims to establish a European Tertiary Education Register to supply the required data (Bonaccorsi et al., 2010; ETER, 2014). Regardless of its likelihood to overtake the 'Big Three', U-Multirank has been broadly welcomed, and its influence is evident in the way other rankings have begun to facilitate personalization of the data.

References

Alperin J. P. (2013) 'Ask Not What Altmetrics Can Do for You, But What Altmetrics Can Do for Developing Countries'. *ASIS&T Bulletin*, May/June, http://www.asis.org/Bulletin/Apr-13/AprMay13_Alperin.html, date accessed 15 April 2014.

ARC (Australian Research Council) (2010) 'Tiers for the Australian Ranking of Journals', http://www.arc.gov.au/era/tiers_ranking.htm, date accessed 28 November 2010.

Arnone M. (2003) 'The Wannabes', *The Chronicle of Higher Education*, 3 January, A18–20.

Bonaccorsi A., Brandt T., De Filippo D., Lepori B., Molinari F., Niederl A., Schmoch U., Schubert T., Slipersaeter S. (2010) *Feasibility Study for Creating a European University Data Collection. Final Study Report. Contract No. RTD/C/C4/2009/0233402* (Brussels: The European Commission), http://ec.europa.eu/research/era/docs/en/eumida-final-report.pdf, date accessed 21 April 2013.

Boulton G., Lucas C. (2008) *What are Universities For?* (Leuven: League of European Research Universities), http://www.leru.org/?cGFnZT00, date accessed 5 May 2010.

Boyer E. L. (1990) *Scholarship Reconsidered. Priorities of the Professoriate* (Princeton, NJ: Carnegie Foundation for the Advancement of Teaching).

Boyer E. (1996) 'The Scholarship of Engagement'. *The Journal of Public Service and Community Outreach*, 1(1) 11–20.

Bush V. (1945) *Science, the Endless Frontier. A Report to the President by Vannevar Bush, Director of the Office of Scientific Research and Development. Nature* (Vol. 188). (Washington D.C.: *United States Government Printing Office*). https://www.nsf.gov/od/lpa/nsf50/vbush1945.htm, date accessed 19 August 2014.

Carayannis E. G., Campbell D. F. J. (2009) '"Mode 3" and "Quadruple Helix": Toward a 21st Century Fractal Innovation Ecosystem'. *International Journal of Technology Management*, 46 (3–4) 201–234.

Castells M. (1994) 'The University System: Engine of Development in the New World Economy'. In J. Salmi, A. M. Vespoor (eds) *Revitalizing Higher Education* (Oxford: Pergamon), 14–40.

CFIR (Committee on Facilitating Interdisciplinary Research) (2004) *Facilitating Interdisciplinary Research* (Washington, DC: The National Academies Press), http://books.nap.edu/openbook.php?record_id=11153&page=2, date accessed 31 July 2009.

Delanty G. (2001) *Challenging Knowledge. The University in the Knowledge Society* (Buckingham: SRHE/Open University Press).

Dempsey N. (2004) 'Minister for Education and Science Address at the Europe of Knowledge 2020 Conference', 24 April 2004, Liege.

Dewatripont M. (2008) 'Reforming Europe's Universities', BEPA [Bureau of European Policy Advisers], Monthly Brief, 14.

Dill D., van Vught F. (2010) 'Introduction'. In D. Dill, F. van Vught (eds) *National Innovation and the Academic Research Enterprise. Public Policy in Global Perspective* (Baltimore, MD: Johns Hopkins University Press), 1–26.

Dunleavy P. (ed) (2013) *Open Access Perspectives in the Humanities and Social Sciences* (London: Sage Publications).

Editorial (2010) 'Save our Cities'. *Nature*, 467 (7318) 883–884.

Ehrenberg R. G. (2002) 'Reaching for the Brass Ring: The U.S. News & World Report Rankings and Competition [Electronic version]'. *Review of Higher Education*, 226 (2) 145–162.

ESF (European Science Foundation) (2010) 'European Reference Index for the Humanities (ERIH)', http://www.esf.org/research-areas/humanities/erih-european-reference-index-for-the-humanities.html, date accessed 1 June 2010.

Espeland W. N., Sauder M. (2007) 'Rankings and Reactivity: How Public Measures Recreate Social Worlds'. *American Journal of Sociology*, 13 (1) 1–40.

ETER (European Tertiary Education Register) (2014) http://eter.joanneum.at/imdas-eter/, accessed 19 August 2014.

Etzkowitz H., Leydesdorff L. (1997) *Universities and the Global Knowledge Economy: A Triple Helix of University-Industry-Government Relations* (Andover: Thomson Learning).

Europa (2000) 'Presidency Conclusions', Lisbon European Council, 23–24 March, http://www.europarl.europa.eu/summits/lis1_En.htm or http://europa.eu/scadplus/glossary/research_and_development_En.htm, date accessed 3 May 2010.

Europa (2010) *ERA in the Knowledge Triangle* (Brussels: European Commission), http://ec.europa.eu/research/era/understanding/what/era_in_the_knowledge_triangle_En.htm, date accessed 14 July 2010.

Europa (2013) *The Grand Challenge. The Design and Societal Impact of Horizon 2020* (Brussels: European Commission).

Garfield E. (1955) 'Citation Indexes to Science: A New Dimension in Documentation Through Association of Ideas'. *Science*, 122 (3159) 108–111.

Garfield E. (2005) 'The Agony and the Ecstasy: The History and Meaning of the Journal Impact Factors', Paper to International Congress on Peer Review and Biomedical Publication, Chicago.

Gibbons M. (2002) 'Engagement as a Core Value in a Mode 2 Society'. In S. Bjarnason, P. Coldstream (eds) The Idea of Engagement. Universities in Society (London: Association of Commonwealth Universities), 48–70.

Gibbons M., Limoges C., Nowotny H., Schwartzman S., Scott P., Trow M. (1994) *The New Production of Knowledge* (London: Sage).
Goddard J. (2009) *Re-inventing the Civic University* (NESTA: London).
Graham W. R. (1987) *A Research and Development Strategy for High Performance Computing* (Washington, D.C.: Office of Science and Technology Policy).
Hazelkorn E. (2009) 'The Impact of Global Rankings on Higher Education Research and the Production of Knowledge', Paper No. 16, UNESCO Forum on Higher Education, Research and Knowledge Occasional, http://unesdoc.unesco.org/images/0018/001816/181653e.pdf, date accessed 3 May 2010.
Hazelkorn E. (2010) 'Community Engagement as Social Innovation'. In L. Weber, J. Duderstadt (eds) *The Role of the Research University in an Innovation-Driven Global Society* (London and Geneva: Economica).
Hazelkorn E. (2011) *Ranking and the Reshaping of Higher Education: The battle for World Class Excellence* (Basingstoke: Palgrave Macmillan).
Hazelkorn E. (2012a) 'European "Transparency Instruments": Driving the Modernisation of European Higher Education'. In P. Scott, A. Curaj, L. Vlăsceanu, L. Wilson (eds) *European Higher Education at the Crossroads: Between the Bologna Process and National Reforms*, vol. 1 (Dordrecht: Springer), 339–360.
Hazelkorn E. (2012b) '"Everyone Wants to Be Like Harvard" – or Do They? Cherishing All Missions Equally'. In P. Scott, A. Curaj, L. Vlăsceanu, L. Wilson (eds) *European Higher Education at the Crossroads: Between the Bologna Process and National Reforms*, vol. 2 (Dordrecht: Springer), 837–862.
Hazelkorn E. (2013a) 'Striving for "World Class Excellence": Rankings and Emerging Societies'. In D. Araya, P. Marber (eds) *Higher Education in the Global Age: Universities, Interconnections and Emerging Societies* (New York: Routledge Studies in Emerging Societies, Routledge), 246–270.
Hazelkorn E. (2013b) 'Higher Education's Future: A New Global Order?' In R. Pritchard, J. E. Karlsen (eds) *Resilient Universities. Confronting Changes in a Challenging World* (Bern: Peter Lang), 53–90.
Hazelkorn E. (2013c) 'Europe Enters the College Ranking Game', *Washington Monthly*, September/October, http://www.washingtonmonthly.com/magazine/september_october_2013/features/europe_Enters_the_college_rank046894.php, accessed 19 August 2014.
Hazelkorn E. (2015 forthcoming) *Ranking and the Reshaping of Higher Education: The Battle for World Class Excellence*, 2nd edn (Basingstoke: Palgrave Macmillan).
Hazelkorn E., Ryan M. (2013) 'The Impact of University Rankings on Higher Education Policy in Europe: A Challenge to Perceived Wisdom and a Stimulus for Change'. In P. Zgaga, U. Teichler, J. Brennan (eds) *The Globalization Challenge for European Higher Education: Convergence and Diversity, Centres and Peripheries* (Bern: Peter Lang), 79–100.
Karabel J. (2005) *The Chosen. The Hidden History of Admission and Exclusion at Harvard, Yale and Princeton* (Boston, MA and New York: Houghton Mifflin Co).
Lederman D. (2005) 'Angling for the Top 20', *Inside Higher Ed*, 6 December 2010. http://www.insidehighered.com/news/2005/12/06/Kentucky, accessed 19 August 2014.
Ludwig M., Scharrer G. (2009) 'Emerging Texas universities look for boost to top tier', *San Antonio Express News*, 30 August. http://www.chron.com/news/houston-texas/article/Emerging-Texas-universities-look-for-boost-to-top-1750387.php, accessed 19 August 2014.

Lundvall B.-Å. (ed.) (1992) *National Systems of Innovation: Towards a Theory of Innovation and Interactive Learning* (London: Pinter Publishers).
Lynton E. A. (1987) *New Priorities for the University: Meeting Society's Needs for Applied Knowledge and Competent Individuals* (San Francisco: Jossey-Bass Higher Education Series, Proquest Info & Learning).
Maslen G. (2013) 'Research Universities to Establish Global Network'. *World University News*, 21 March, http://www.universityworldnews.com/article.php?story=20130321080356646, date accessed 24 March 2013.
Mohrman K., Ma W., Baker D. (2008) 'The Research University in Transition: The Emerging Global Model'. *Higher Education Policy*, 21 (1) 29–48.
National Academy of Sciences (2004) *Facilitating Interdisciplinary Research*, Committee on Facilitating Interdisciplinary Research, *National Academy of Sciences, National Academy of Engineering, and Institute of Medicine of the National Academies* (Washington, D.C.: The National Academies Press).
Nelson R. (1993) *National Innovation Systems: A Comparative Analysis* (Oxford: Oxford University Press).
O'Connor J. (1973) *Fiscal Crisis of the State* (New York: St Martin's Press).
O'Meara K., Rice E. R. (eds) (2005) *Faculty Priorities Reconsidered: Rewarding Multiple Forms of Scholarship* (San Francisco: John Wiley & Sons).
OECD (2002) *Frascati Manual 2002. Proposed Standard Practice for Surveys on Research and Experimental Development*, 6th edn (Paris: Organisation of Economic Co-operation and Development).
OECD (2009) *Education at a Glance* (Paris: Organisation of Economic Co-operation and Development).
OECD Education Ministerial Meeting (2010) 'Investing in Human and Social Capital: New Challenges' (Paris: Organisation of Economic Co-operation and Development), http://www.oecd.org/dataoecd/53/16/46335575.pdf, date accessed 2 December 2010.
Porter M. (1985) *Competitive Advantage* (New York: Free Press).
Reichert S. (2009) *Institutional Diversity in European Higher Education. Tensions and Challenges for Policy Makers and Institutional Leaders* (Brussels: European University Association).
Rothwell R. (1994) 'Towards the Fifth-generation Innovation Process'. *International Marketing Review*, 11 (1) 7–31.
Salmi J., Saroyan A. (2007) 'League Tables as Policy Instruments: Uses and Misuses'. *Higher Education Management and Policy*, 19 (2) 31–68.
Saltmarsh J., Hartley M., Clayton P. (2009) 'Democratic Engagement White Paper', New England Resource Center for Higher Education Publications. Paper 45 (University of Massachusetts, Boston), http://scholarworks.umb.edu/nerche_pubs/4, accessed 19 August 2014.
Terenzini P. T., Ro H. K., Yin A. C. (2010) 'Between-College Effects on Students Reconsidered', paper presented at the Meeting of the Association for the Study of Higher Education, Indianapolis, IN.
Tierney W. G. (2009) 'Globalization, International Rankings, and the American Model: A Reassessment'. *Higher Education Forum*, 6, 1–18.
Trow M. (1974) 'Problems in the Transition from Elite to Mass Higher Education', *General Report on the Conference on Future Structures of Post-Secondary Education*, (Paris, OECD), 55–101. (Reprinted in M. Burrage (ed.) (2010) *Martin Trow.*

Twentieth-Century Higher Education. From Elite to Mass to Universal (Baltimore, MA: Johns Hopkins University Press, 88–143).

Usher A., Jarvey P. (2010) 'Let the Sun Shine In: The Use of University Rankings in Low- and Middle-income Countries', paper presented at the IREG 5 Conference, Berlin.

Usher A., Medow J. (2009) 'A Global Survey of University Rankings and League Tables'. In B. M. Kehm, B. Stensaker (eds) *University Rankings, Diversity, and the New Landscape of Higher Education* (Rotterdam: Sense Publishers), 3–18.

Van Vught F. V., Ziegele F. (eds) (2011) *Design and Testing the Feasibility of a Multidimensional Global University Ranking Final Report*. http://ec.europa.eu/education/library/study/2011/multirank_En.pdf, accessed 19 August 2014.

Van Vught F. A., Ziegele F. (eds) (2012) *Multidimensional Ranking. The Design and Development of U-Multirank* (Dordrecht: Springer Publishers).

Ward E., Hazelkorn E. (2012) 'Engaging With The Community'. In S. Bergan, E. Egron-Polak, J. Kohler, L. Purser, M. Vukasović (eds) *Handbook on Leadership and Governance in Higher Education* (Stuttgart: Raabe Verlag).

Webster D. S. A. (1986) *Academic Quality Rankings of American Colleges and Universities* (Springfield: Charles C. Thomas).

Webster T. J. (2001) 'A Principal Component Analysis of the U.S. News & World Report Tier Rankings of Colleges and Universities'. *Economics of Education Review*, 20, 235–244.

World Bank (2009) 'Averting a Human Crisis During the Global Downturn', Policy Options from the World Bank's Human Development Network Conference, Washington, DC. http://siteresources.worldbank.org/NEWS/Resources/AvertingTheHumanCrisis.pdf, accessed 19 August 2014.

World Health Organization (2009) 'Third High-level Preparatory Meeting', Bonn, Germany, 27–29 April 2009, http://www.euro.who.int/__data/assets/pdf_file/0013/105160/3rd_prep_mtg_bonn.pdf, date accessed 18 April 2014.

3
Linking Quality Assurance and ESD: Towards a Participative Quality Culture of Sustainable Development in Higher Education

Oliver Vettori and Christian Rammel

Introduction: sustaining sustainability

Education for sustainable development (ESD) is nothing less than a radical paradigmatic shift in education and learning. To initiate this shift, the United Nations Decade of Education for Sustainable Development (UN DESD) has the overall objective to integrate the main principles and values that underlie sustainable development into all aspects of learning and education (UNESCO, 2005). This global educational effort is intended to encourage far-reaching changes in behaviour that will create a more sustainable future and will enable citizens to tackle both present and future sustainability challenges. In short, ESD shall act as a fundamental agent of sustainable change.

Higher education institutions have a highly relevant role in contributing to ESD through the generation of knowledge related to sustainability challenges and the competences to tackle these challenges (UNECE, 2005). Moreover, they are challenged to cooperate in networks that constitute a supportive infrastructure for ESD processes as well as to develop their potential to become active nodes between formal, non-formal and informal educational partners in order to initiate and enhance stakeholder dialogue via science–society interfaces (van Dam-Mieras, 2006). Authors such as Scott et al. (2012) go even further and propose a clear leadership role for higher education in the field of ESD, as universities shape the future leaders and decision-makers who will have to address the key problems of our world.

Yet, in 2014, the UN Decade of Education for Sustainable Development will end and a new UN decade will shift our focus towards other issues on a global scale. Nearing the end of the current decade, we are increasingly faced with determining if education has really become a major driving force towards sustainable change, if our education systems have changed in a way that supports transformative learning and if the previous years have indeed given birth to a paradigmatic shift in education reflected by developments such as second order learning, reflectivity or participatory teaching.

Looking at the still growing number of examples of successfully implemented sustainability processes and projects in higher education institutions, we would be tempted at first to give a positive answer. The 'greening the campus' wave that had started in the USA in the past years has already reached Europe and is a huge influence on new campus development projects. There are more and more 'sustainability offices' at universities worldwide (Leal Filho, 2011) and new international treaties such as the 'People's Sustainability Treaty on Higher Education' (IUCN, 2012), the 'Higher Education Sustainability Initiative for Rio+20' (UNCSD, 2012) or the '50+20 Agenda' (50+20, 2012), show that sustainability is an issue that has attracted more and more attention.

Nevertheless, gaining more attention in policy documents, the creation of sustainability offices or 'greening' the campus is not enough to mean a radical transformation in higher education. This requires a clear reorientation of existing curricula, management, research and pedagogy towards the objectives of the UN DESD. Far from simply adapting courses or the content of educational structures, ESD demands a holistic approach,[1] which is expressed by the transformation of the way universities are operating and structured (Scott et al., 2012). At the end of the UN DESD, higher education pioneers who adopt an integrative 'whole institution approach' in ESD such as Leuphana University (Germany) or the University of Florida (USA) are still the exception rather than the rule. In order to avoid what happened in the follow-up phase after Rio 1992, one thing seems to be the imperative: in order to become sustainable itself, sustainable development in higher education needs to become a fully integrated part of education, starting at the level of the individual education institution (UNECE, 2005). If sustainable development remains just an 'add on' for education institutions and is mainly dependent on individual pioneers, its objectives will not be reached and it will disappear in the continuous struggle between the emergence of new trends in other areas and budgetary restrictions.

In this chapter, we will argue that a particularly promising strategy to achieve the goal of integrating ESD into the core activities of universities lies in linking sustainability issues to an institution's quality culture and the institutional quality management systems that build upon it. With regard to the value-bound and context-dependent character of educational quality per se, sustainable development can be viewed as an operational quality goal, but also as an objective and key value for the development of the institutional quality culture itself. In this regard, the benefit could well be mutual: showing an overwhelming preoccupation with standards and procedures, many quality assurance (QA) activities seem to be in need of a clearly defined purpose (Williams, 2009), which the reference to sustainability could provide. We will further develop this line of argument in the course of this chapter.

The next section will first give an overview of problems in the current international QA discourse that are also relevant for the topic of ESD. In a second section, we will introduce the concept of an institutional quality culture as a means of engaging the institutional community as a whole in a common value-driven purpose. The chapter will conclude with some suggested ways of engaging the members and stakeholders of an educational institution in a sustainability-oriented quality culture. By taking an organization culture perspective, we will show how new tools and initiatives need to be related to already existing structures of action and patterns of interpretation in order to become an integral part of organizational life. Particular attention will be paid to the question how an organization's 'communication architecture' should be built in order to allow the various actors to participate instead of merely enacting (and potentially rejecting) an externally imposed 'script'.

Quality assurance in higher education and the problem of engaging the key stakeholder

Participation is one of the key concepts ESD is built on. In fact, ESD shifts the focus away from solely transmitting knowledge and information towards participatory learning and is also expected to enhance participatory democracy (UNECE, 2005; Thessaloniki Declaration, 1997). Herein, a basic intention of participatory ESD is to empower people through education so that they can actively shape sustainable development processes. However, engaging stakeholders within ESD processes is still difficult to achieve. Especially in the context of educational institutions, we can identify a long-lasting debate about top-down versus bottom-up participation and the need to contest people's narrow personal interests when it comes to achieving sustainability within a community or institution (Laessoe, 2007 and 2010). Therefore, authors such as Laessoe (2010) argue for developing a new approach to participatory ESD which works with dilemmas and dissent to enhance a critical and creative process of engaging people.

The problem of engaging the entire institutional community is also one of the key issues at the heart of the professional quality assurance discourse – at least in the European context, where the Berlin Communiqué in 2003 did not only place quality 'to be at the heart of a European Higher Education area', but also stated that the primary responsibility for quality lies with the institution itself (EHEA, 2003).

But in order to understand where the engagement issue stems from, it might be helpful to provide some historical context (Vettori, 2012a). Even though the roots of quality assurance as a policy field can be traced back to the end of the 19th century and the emergence of the first accreditation agencies in the United States, it was not until the effects of the transition from 'elite' to 'mass' higher education became fully apparent in the early 1980s that policy- and decision-makers started to develop the first formal quality assurance schemes (Westerheijden et al., 2007). Until the 1970s, quality in higher education was very much 'influenced' rather than 'managed' or 'controlled'. This early form of quality assurance was achieved through a variety of bureaucratic means, mostly focusing on the input level (e.g. legal conditions, state-provided funding tied to formalized rules). From the late 1970s, quality assurance was emerging more and more as a phenomenon of its own, reflecting the increasing importance of extrinsic values in higher education (Van Vught, 2000) and mimicking the development in the more industrialized sectors (Schwarz and Westerheijden, 2004). The 1990s saw the implementation of formal quality assurance instruments and processes (e.g. self-assessment, supporting documentation, peer review or public reports) in most European countries (Harvey, 2006), but also overseas (Tam, 1999). The internal mechanisms and instruments were at the same time complemented by the introduction of national quality assurance systems and a growing focus on external quality assurance. In 2008, an OECD publication declared the development of external quality assurance systems as one of the most important trends in higher education in the last decades (Riegler, 2010, p. 157).

However, among some of the key actors in higher education, these developments were far from being welcomed: From the very beginning, many quality assurance approaches were criticized as being overly managerialistic and formalistic and thereby met with a distinct lack of enthusiasm from most academics (Newton, 2000 and 2002; Anderson, 2006). Quality assurance was perceived as an externally imposed burden that was more about window-dressing and 'feeding the beast' (Newton, 2002) of bureaucracy than about achieving the kind of excellence in teaching or even 'transformative learning' (Harvey and Knight, 1996) the approaches were supposedly aiming for. As a result, quality assurance was delegated to special quality assurance units, and the question of how to engage students and academics in developing the institutional quality assurance systems – and, more importantly, filling them with life and meaning – remains largely unsolved. Much like with sustainability, the problem is not that actors object to the key issue as such – for who would argue against quality and improvement – but that they apparently do not take the issue further than is absolutely necessary. The problem seems to be, how to transform this abstract acceptance of a value into the actors' daily practices and interactions. In other words: make it a part of the institutional culture.

Engagement and participation as key objectives: the quality culture movement

In order to address the problem, newer quality assurance approaches such as the quality culture model introduced by the European University Association (EUA, 2006) aim to shift their focus to more development-oriented and value-based aspects (Vettori, 2012a). In such a perspective, formal quality assurance systems are complemented by a 'softer' cultural side based on values and practices that are shared by the institutional community and that have to be nurtured on many levels and by various means at the same time (EUA, 2005 and 2006). In essence, the concept tries to reframe quality assurance as a core value of higher education institutions instead of an externally imposed chore: 'A culture of quality is one in which everybody in the organisation, not just the quality controllers, is responsible for quality' (Crosby, 1986 cited in Harvey and Green, 1993, p. 16). In this regard, the quality culture concept is putting a stronger focus on the behaviour of stakeholders rather than on the operation of a quality system (Harvey, 2007, p. 81). Following Edgar Schein, Ehlers (Ehlers, 2009) regards quality culture as an answer to the question in which way an organization is responding to its quality challenges and is fulfilling its quality purpose.

In EUA's quality culture perspective, quality is not beheld as a process that can be operated through evaluation and measurement procedures alone, but as values and practices that are shared by the institutional community and that have to be nurtured on many levels (e.g. by considering the sub-cultures in the respective academic subunits) and by various means at the same time. The approach demands the involvement of multiple internal and external stakeholders, acknowledging the fact that a quality culture cannot be implemented from above, although strong leadership may be necessary for starting and promoting the process in the first place. Quality measurement and quality control are undoubtedly important elements of such an approach (as they are of any quality management system), but they cannot be regarded as quality guarantors per se, rather needing to be embedded in an overarching framework that is in line with the institutional objectives and focuses on continuous improvement (Vettori et al., 2007, p. 22).

In recent years, the concept has also received more and more scholarly attention (e.g. Lueger and Vettori, 2007; Vettori et al., 2007; Harvey and Stensaker, 2008; Ehlers, 2009; Harvey, 2009). In general, the academic discussion is very much mirroring the debate on organizational culture since the 1980s, contrasting a functionalist approach and a rather interpretative, interactionist/socio-constructivist approach. Within the functionalist group, culture is seen as something an organization has, that is, culture as a potentially identifiable and manipulative factor. From an interpretative viewpoint, culture can be regarded as something an organization is, that is, culture as an integrated product of social interaction and organizational life impossible to differentiate from other factors (Harvey and Stensaker 2008, p. 431). In this chapter, we take a rather interactionist view on the phenomenon, with a quality culture being defined as 'a socially mediated and negotiated phenomenon leading to shared results of meaning construction which is largely unconscious and only in some elements directly visible to the outside' (Ehlers, 2009, p. 352). Here, different understandings of quality are already embedded in several contexts, of which an organization's culture is one of the most important ones (Vettori, 2012a). From such a perspective, culture is not fixed and stable, but can be regarded as the result of multiple interactions, involving all participants of these interactions (Smircich, 1983 and 1985; Allaire and Firsirotu, 1984; Weick, 1994; or Froschauer, 1997).

As a consequence, the challenge for any university management lies in creating a setting that is helping to foster an internal quality culture, not in managing this culture. If the university's management tries to dominate the inner-institutional quality discourse and aims at controlling the respective development processes, this may very well backfire and lead to a completely different kind of quality culture than the intended one: quality-related activities might quickly degrade to a mere schematism, which other stakeholders (especially the academic units) try to subvert or simply ignore. If a quality culture should indeed be sustained by the whole organization, its basic principles have to be largely shared or at least accepted. Such expectations presuppose a considerable amount of participatory involvement and trust. Key factors in this regard are the patterns of explanation and interpretation that prevail in a certain organizational context. Influencing these patterns can – at least in a long-term perspective – produce better results than simply reworking evaluation methods or implementing new procedures (even though both sides are obviously interrelated). On that score, assisting university members with interpreting their daily actions from a certain perspective – for example, sustainable development – and empowering their autonomous decision-making (but also to understand and accept the consequences of such decisions) are important aspects of a functioning quality culture.

Linking quality cultures and ESD

At first glance, the differences between ESD and QA might seem bigger than the similarities: the two issues have quite different origins, they are perceived in very different ways within the higher education field and they are oriented towards different functions and institutional/societal objectives. On the other hand, their strong connection to individual and collective values and the necessity to become an integrated part of all actors' daily actions and interactions in order to achieve the intended effects, can also be regarded as a strong connection between the two issues. In addition, ESD carries its purpose very much on its sleeve: in fact, sustainability as such is, at least in theory, a goal that seems to be unanimously acceptable if not even desirable. Arguably, the same could be said about quality improvement, yet considering the relativity of the quality concept, it rather seems as an empty canvas: a concept that is bare of any inherent meaning which invites a great diversity of concepts to be projected onto it (Vettori, 2012a). The struggle or search for a unifying purpose seems one of the bigger problems in contemporary quality assurance (Vettori, 2012a). Putting the existing QA structures – and, as we argue in this chapter, the quality culture approach – at least partly into the service of ESD, could therefore well be of mutual benefit for the two issues/communities.

In the following paragraphs, we will show three important links between ESD and QC: (1) values, (2) multidimensional quality concepts and (3) stakeholder engagement/participation as central features. All three dimensions are not only strong conceptual links between the two constructs but can also be regarded as starting points for future developmental actions: an argument that will be further developed in the final section.

As we have already discussed, the focus on individual and organizational values is one of the key characteristics and strengths of the quality culture approach. The focus on values is also at the very heart of ESD as the entire concept aims to integrate the fundamental values that underlie sustainable development into all aspects of learning and education. Moreover, understanding our own values, the values of the society we live in and the different values of other people and cultures is crucial for educating others and ourselves for a sustainable future. Accordingly, the international implementation scheme of the UN DESD states: 'Each nation, cultural group, and individual must learn the skills of recognizing their own values and assessing these values in the context of sustainability' (UNESCO, 2005, p. 7). In principle, the values of sustainable development are related to human dignity, rights, equity and care for the environment, but are notably extended beyond the present generation towards the impact on future generations. To list only a few central ones, sustainable development values bio- and cultural diversity, inclusivity, democracy, participation, self-sufficiency and equity. The concrete values the ESD programme should promote and put forward is still a matter of discussion as different values are linked with different legitimate perspectives. This diversity of values and interests is also one of the core areas the quality culture approach focuses on, emphasizing the necessity to critically reflect value conflicts and utilize them in a productive manner: Shared values may be an important basis to work from, but the strength of the approach lies also in accepting and actively using culturally induced conflicts as a driver for constructive change.

The issue of quality in education has a long-standing history (Ross and Genevois, 2006). Recently, quality is more and more understood as a dynamic concept that has to constantly adapt to a changing world, where societies are undergoing fundamental socio-economical transitions. Regarding the various legitimated perspectives and topics which drive these transitions, there is a growing awareness to understand ESD more in the light of a guiding quality goal itself which directs the educational process towards enabling learners to make decisions that enhance and shape a sustainable development, and ais thus an imperative requirement for building sustainability in the future (Pigozzi, 2007; Mulà and Tilbury, 2009). It is also an explicit goal of the UN DESD to link the quality education agenda to sustainability matters (Hopkins, 2007). In fact, ESD is expected to become an integrative part of the various aspects of a quality education and not be seen as only a stand-alone subject. To go even one step further, if ESD is understood as a meaningful complementary part of quality education, sustainability can play a significant role in national lifelong learning strategies as well as be used as a tool to enhance quality at all levels of education and training.

Similar to ESD, the understanding of 'quality' in the quality culture approach is quite a dynamic one. In both areas, quality is a process that is based more on values, practices and continual interactions within the institutional community rather than managed and controlled by evaluation and measurement procedures. Until recently, the traditional closeness between quality assurance and performance management has limited the integration of ESD (Martin et al., 2009). If the focus is primarily on meeting the objectives of the next inspection or quality audit, there is no room for the dynamic and long-term horizon of ESD, which seeks to enable practitioners and learners to explore a more transformative educational experience. But as the quality culture approach goes far beyond performance management or audits and has disrupted the narrowly prescriptive framework of quality management, there is now room to integrate sustainable development as a cross-sectional issue in areas such as curricula, management, teaching or institutional decision-making.

Last but not least, participation/stakeholder engagement is described as one of the central aspects if we want to translate visions of sustainability into praxis (Laesso, 2010). The diversity of perspectives and values inherent in sustainable development means that there will (and must) always be multiple entry points for participation in general ESD processes. Hereby, an added value of ESD is that it recognizes that these perspectives are inter-linked and that it provides a framework where these interests can collectively shape the common endeavour to educate and learn for a sustainable future. As ESD aims at the generation of a culture of mutual respect and learning in decision-making and communication, it shifts also the focus away from transmitting information towards participatory learning. Participatory learning does not only improve the quality of decision-making, but also effectively resolves conflict among competing interests, builds trust in institutions, and educates and informs related stakeholders (UNECE, 2005). Therefore, many crucial aspects of ESD processes such as science–society interfaces, adaptive learning communities or community-based decision-making are based on the simple precondition to engage and integrate stakeholders from the beginning. Additionally, as ESD contributes to achieving justice and resolving societal conflicts, it simultaneously enhances participatory democracy.

Similar to the quality culture approach, the main goal of the engagement principle is to create a sense of ownership for the quality improvement/sustainability processes, not only on the individual but on the institutional level. In analogy to a quality culture where everybody in the university is responsible for developing the quality of its products, structures and processes, ESD aims at a culture where every actor takes responsibility for sustainability as a specific quality issue. In participative ESD, it is essential to maintain stakeholder involvement over time, which is only possible if the stakeholders find the proper setting where they are able and willing to communicate, learn and engage. Therefore, a well-functioning and dynamic quality culture does not only provide the space for the involvement of multiple internal and external stakeholders, it also offers the room for keeping these stakeholders 'busy' – 'busy' in the sense that keeping the discourse alive and running can be regarded as an important objective per se.

Towards a participative culture of sustainable development

In our previous section, we have shown how ESD could and maybe even should be linked to existing concepts and initiatives such as the rapidly emerging European quality culture initiative. Yet even if it might not always be possible to form a direct connection or even make ESD an integrate part of an institution's quality assurance system (there might be organizational reasons for this, as well as financial reasons; or the QA system might be under enough pressure from external examiners as it is, without adding another burden), we argue that the cultural perspective as such contains a variety of elements and principles that could be adopted and adapted in order to create sustainable engagement for ESD. In other words: help to develop a participative culture of sustainable development.

In the following paragraphs we will present a few tentative suggestions of how higher education institutions could get there. This is far from being a full-fledged and entirely consistent framework though, but rather a first collection of ideas and steps that could be taken in order to start moving in this direction.

Use the quality culture concept as a tool for reflection: Taking the quality culture concept as a purely normative ideal can have some severe drawbacks (Vettori, 2012b). As with any construct that is bound to values, it is highly unlikely that the same normative ideal would appeal to every institution and actor. In other words: something that is viewed as an improvement from a student perspective can be regarded as a change for the worse from a teacher's perspective and vice versa. With regard to the complexities involved in sustainability issues, this becomes even trickier. Hence, if the goal is to achieve a higher level of acceptance and participation, imposing a supposedly 'good' idea on others can be quite risky, particularly if the idea gets subverted in practice. In addition, the framing of culture as a potentially identifiable and manipulative factor among others also means to underestimate the importance of the value and belief systems that underlie all organizational activities, events or observations (Vettori, 2012b, p. 5). Taking this into account results in a reframing of the concept in the interactionist tradition described above: then, a quality culture is not a universally shared ideal nor does it even provide a common goal for each institution, but rather a common starting point. By using the concept as an analytical tool (Harvey and Stensaker, 2008) rather than as a set of procedures (Harvey, 2009), the concept allows its users to reflect current strategies, practices and principles with regard to their cultural resonance and relevance. Following this logic leads to questions such as: What is important and from whose point of view? Is there a gap between the rhetorics and reality? Are the institutional identity and values mirrored in the strategic documents? How well does the institutional culture deal with change? Do the institutional strategies provide for the possibility (and necessity) of change in the institution and the institutional environment? How and how well can it cope with conflicts that may arise? Are the current tools and practices finding resonance with the stakeholders and actors they are intended to benefit? If not, why not? Which are the main inconsistencies and contradictions in and between the institution's sustainability rhetorics and actions? Such an approach might not only give the formulation of ESD strategic documents a different spin, but can also beware the respective policies from promoting a logic that is incompatible with the organization's current cultural imperatives.

Create a sense of ownership: Adherence to an empowerment approach implies to make productive use of the actors' self-organizational abilities (Fetterman, 2001). If, as we have argued above, constructive stakeholder engagement goes beyond acting out an externally imposed script, a quality culture perspective might well require that the different actors are challenged as well as enabled to contribute to the script themselves: to formulate their own goals and develop their own activities within a shared framework. In simpler words: within the quality culture approach shared responsibilities also means shared opportunities and risks. With higher education institutions usually consisting of highly individualized and loosely coupled experts (Pellert, 1999), it seems even more relevant to encourage the various actors to try out their own ideas and approaches – and to participate in the consequences as well, independent of their success. On the other hand, decision-makers will be confronted with the challenge to hand over some of their competences and to pull at least partly out of the operational implementation process. This might prove even more difficult, if the initiative had been started top-down and if the university management regards itself not as a mere promoter of the process, but as its main mover. Within a stakeholder-oriented approach, the university management represents just one of various perspectives, although with an arguably higher probability of being asserted (Vettori et al., 2007, p. 24). On the other hand it is not sufficient to grant responsibility and autonomy: the process participants should feel compelled to actually take it. Communication and mutual trust can form the basis for this interplay of delegation and acceptance, but they depend on a successful implementation of trust-building actions.

Develop a communication architecture that fosters dialogue and discussion: In many ways, organizations rely on communication. The key role of communicative acts is well accepted in QA as well as in the ESD discourse, yet in our view the dominating communication model is still one of a mechanism for transmitting information such as in the Shannon-Weaver-Model (Weaver and Shannon 1963), rather than understanding communication as a process for generating and sharing meaning. From the latter perspective it quickly becomes clear why information is rarely interpreted in the way the communicator intends it to be. Even communication channels are usually imbued with meaning and are often dealt with accordingly: using a committee structure for dealing with sustainability makes it 'official' but also frames it as a part of the usual political negotiations that are a major characteristic of committees. Putting the latest sustainability activities in the official institutional newsletter might stir the interest of external stakeholders, but can also lead to the internal view that this is just 'another marketing trend'. Last but not least, the meaning of a certain phrase or term can go far beyond the obvious: the idea of continuous improvement, for example, seems to be a universally acceptable one and can be found across a great number of higher education policies and strategies. In combination with an output logic, however (e.g. by framing high quality as an increase of high-impact publications or constantly higher student satisfaction results), the ideal might well transform into a demotivating risk, signalling that, no matter how much people invest, it will never be enough (Vettori, 2012b). Consequently, within a quality culture framework, it needs to be taken into account that communications can hardly be managed or controlled – but they can be fostered in creative ways. Discussion and negotiation formats play a pivotal role here, allowing a variety of actors to construct the communicative reality of sustainability together instead of receiving information and acting accordingly to the normative layer encoded in this information. Shared constructions are not as frequent as we may think – particularly when it comes to abstract concepts such as sustainability or quality – but have to be carefully developed. In addition a certain term or phrase that is central for the issue as such can 'wear out', particularly if it has been associated with specific actors or socio-historical contexts. By a continuous 'reframing' of central issues (e.g. changing key words and terms every few years while staying true to the underlying principle), the discourse can be kept fresh – as long as the new frames find resonance in the already existing organization culture.

In this respect, it might also be necessary for key actors such as sustainability managers or promoters to function as interpreters and 'cultural brokers' between different languages and institutional sub-cultures. Overall, institutional sustainability managers have a difficult task: they need to promote the overall awareness and development process as well as initiate and coordinate activities, but at the same time must avoid that responsibility is entirely shifted to them. The better they do their jobs, the more the organization will start to depend on them, but this might also lead to a decrease of others' engagement, choosing the most resource-efficient way of achieving the overall objectives. Yet considering the diversity of perspectives and potentially conflicting value patterns that characterize higher education institutions as much as the sustainability discourse, the most challenging (and arguably most important) duty of such actors might be to foster and moderate the negotiation processes and discussions mentioned above. This is not just a question of social skills, but of being able to communicate across disciplinary borders and organizational hierarchies: a sort of 'intercultural competence' that goes far beyond the capabilities to deal with different countries of origin.

Summing up, we can see that an interactionist cultural perspective can provide some fruitful impulses for strengthening the ESD movement. On the other hand, this chapter can only serve as a starting point for future discussions: many conceptual questions are still open, and the transfer to practice remains another matter altogether. On the other hand, considering that stakeholder engagement is a condition sine qua non for both sustainability and quality assurance, the quality culture approach offers some relevant – and even practical – options for encouraging and increasing stakeholder participation without succumbing to a socio-technical approach of compelling or manipulating people into taking part.

We have started this chapter by pointing out that the United Nations Decade of Education for Sustainable Development is heading towards its end. By forming connections to already existing structures and cultural elements, however, the concept has a good chance to become sustainable itself – maybe not in name but in its core principles. Reframing it as part of an institution's overall quality goals and procedures might not only contribute to achieving a higher level of long-term stability, but could also provide additional directions and means to get there. As we have argued above, a closer connection between ESD and QA – via the quality culture concept – might well be of mutual benefit to both concepts/movements, giving ESD a stable structure to build upon and offering the institutional quality assurance and quality management efforts a constructive orientation beyond the generation of data and evaluation cycles.

Note

1. A 'whole institution approach' means that all aspects of an institution's internal operations and external relationships are reviewed and revised in light of SD principles. Within such an approach each institution would decide upon its own actions addressing the three overlapping spheres of Campus (management operations), Curriculum, and Community (external relationships) (UNECE, 2005).

References

Allaire Y., Firsirotu M. E. (1984) 'Theories of Organizational Culture'. *Organization Studies*, 5 (3) 193–226.

Anderson G. (2006) 'Assuring Quality/Resisting Quality Assurance: Academics' Responses to "Quality" in some Australian Universities'. *Quality in Higher Education*, 12 (2) 161–173.

EHEA (2003) 'Realising the European Higher Education Area', Communiqué of the conference of ministers responsible for higher education in Berlin on 19 September 2003.

Ehlers U. D. (2009) 'Understanding Quality Culture'. *Quality Assurance in Education*, 17 (4) 343–363.

European University Association (EUA) (2005) *Developing an Internal Quality Culture in European Universities. Report on the Quality Culture Project 2002–2003* (Brussels: European University Association).

European University Association (EUA) (2006) *Quality Culture in European Universities: A Bottom-up Approach. Report on the Three Rounds of the Quality Culture Project 2002–2006* (Brussels: European University Association).

Fetterman D. M. (2001) *Foundations of Empowerment Evaluation* (Thousand Oaks, CA: Sage).
Froschauer U. (1997) 'Organisationskultur als soziale Konstruktion'. *Österreichische Zeitschrift für Soziologie*, (2) 107–124.
Harvey L. (2006) 'Understanding Quality'. In L. Purser (ed.) *EUA Bologna Handbook: Making Bologna Work* (Brussels: European University Association and Berlin: Raabe).
Harvey L. (2007) 'The Epistemology of Quality', paper presented at the 8th Biennial Conference of the International Network of Quality Assurance Agencies in Higher Education, Toronto, Canada, 2–5 April.
Harvey L., Green D. (1993) 'Defining Quality'. *Assessment & Evaluation in Higher Education*, 18 (1) 9–33.
Harvey L., Knight P. (1996) *Transforming Higher Education* (Buckingham: Open University Press).
Harvey L., Stensaker B. (2008) 'Quality Culture: Understandings, Boundaries and Linkages'. *European Journal of Education*, 43 (4) 427–441.
Hopkins D. (2007) *Every School a Great School* (Maidenhead, Berkshire: Open University Press, McGraw Hill).
International Union for Conservation of Nature (IUCN) (2012) 'Rio+20 People's Sustainability Treaty on Higher Education', http://www.iucn.org/about/union/commissions/cec/?10113/Rio20-Peoples-Sustainability-Treaty-on-Higher-Education. date accessed 16 July 2014.
Laessoe J. (2007) 'Participation and Sustainable Development – the Post-Ecologist Transformation of Citizen Involvement in Denmark'. *Environmental Politics*, 16 (2) 231–250.
Laessoe J. (2010) 'Education for Sustainable Development, Participation and Socio-cultural Change'. *Environmental Education Research*, 16 (1) 39–57.
Leal Filho W. (ed) (2011) *Sustainable Development at Universities: New Horizons* (Frankfurt: Lang).
Lueger M., Vettori O. (2007) 'Finding the Right Measure? An Interactionist View on Quality Cultures and the Role of Quality Measurement', paper presented at the 8th Biennial Conference of the International Network of Quality Assurance Agencies in Higher Education, Toronto, Canada, 2–5 April.
Martin S., Jucker R., Martin M. (2009) 'Quality and Education for Sustainable Development, Current Content and Future Opportunities'. In L. E. Kattington (2010) *Handbook of Curriculum Development* (New York: Nova Science).
Mulà I., Tilbury D. (2009) 'A United Nations Decade of Education for Sustainable Development (2005–2014): What Difference Will it Make?' *Journal of Education for Sustainable Development*, 3 (1) 101–111.
Newton J. (2000) 'Feeding the Beast or Improving Quality? Academics' Perceptions of Quality Assurance and Quality Monitoring'. *Quality in Higher Education*, 6 (2) 153–163.
Newton J. (2002) 'Views from Below: Academics Coping with Quality'. *Quality in Higher Education*, 8 (1) 39–61.
Pellert A. (1999) *Die Universität als Organisation: Die Kunst, Experten zu managen* (Wien: Böhlau).
Pigozzi M. (2007) 'Quality in Education Defines ESD'. *Journal of Education for Sustainable Development*, 1 (1) 27–35.

Riegler K. (2010) 'Qualitätssicherung: Unde venis et quo vadis? Zur Genese und zukünftigen Entwicklung eines Leitmotivs der europäischen Hochschulreformen'. *Zeitschrift für Hochschulrecht, Hochschulmanagement und Hochschulpolitik*, 9 (6) 157–167.

Ross K., Genevois I. (2006) *Cross National Studies of the Quality of Education: Planning their Design and Managing their Impact* (Paris: UNESCO/IIEP).

Schwarz S., Westerheijden D. F. (2004) 'Accredition in the Framework of Evaluation Activities: a Comparative Study in the European Higher Education Area'. In S. Schwarz, D. F. Westerheijden (eds) *Accreditation and Evaluation in the European Higher Education Area*, Higher Education Dynamics, vol. 5 (Dordrecht: Kluwer), 1–41.

Scott G., Tilbury D. Sharp L., Deane E. (2012) *Turnaround Leadership for Sustainability in Higher Education. Final Report 2012* (Sydney: Office of Learning and Teaching – Department of Industry, Innovation, Science, Research and Tertiary Education).

Smircich L. (1983) 'Organizations as Shared Meanings'. In L. R. Pondy, P. J. Frost, G. Morgan, T. C. Dandridge (eds) *Organizational Symbolism* (Greenwich: Jai Press), 55–65.

Smircich L. (1985) 'Is the Concept of Culture a Paradigm for Understanding Organizations and Ourselves?' In P. J. Frost, L. F. Moore, L. M. Reis, C. C. Lundberg, J. Martin (eds) *Organizational Culture* (Newbury Park: Sage), 55–72.

Tam M. (1999) 'Quality Assurance Policies in Higher Education in Hong Kong'. *Journal of Higher Education Policy and Management*, 21 (2) 215–226.

Thessaloniki Declaration (1997) http://archive.www.iau-aiu.net/sd/rtf/sd_dthessaloniki.rtf, date accessed 30 April 2014.

UNCSD (2012) http://sustainabledevelopment.un.org/index.php?menu=1073.

UNECE (2005) *UNECE Strategy for Education for Sustainable Development* (Vilnius: High-level meeting of Environment and Education Ministries), http://www.unece.org/environmental-policy/areas-of-work/education-for-sustainable-development-esd/about-us/the-strategy.html.

UNESCO (2005) *United Nations Decade of Education for Sustainable Development (2005–2014): International Implementation Scheme* (Paris: Section for Education for Sustainable Development [ED/PEQ/ESD]).

van Dam-Mieras R. (2006) 'Learning for Sustainable Development: Is it Possible Within the Established Higher Education Structures?' In J. Holmberg, B. Samuelsson (eds) *Drivers and Barriers for Implementing Sustainable Development in Higher Education*, Technical Paper No. 3 (Paris: UNESCO), 13–18.

Van Vught F. A. (2000) 'In Search of Quality Management in Western European Higher Education'. In St. Laske, M. Habersam, E. Kappeler (eds) *Qualitätsentwicklung in Universitäten. Konzepte, Prozesse, Wirkungen*, Schriften zur Universitätsentwicklung, Bd. 2 (München und Mering: Rainer Hampp Verlag), 65–99.

Vettori O. (2012a) *A Clash of Quality Cultures? Conflicting and Coalescing Interpretive Patterns in Austrian Higher Education* (Vienna: University of Vienna).

Vettori O. (2012b) *Examining Quality Culture: Part 3 – From Self-reflection to Enhancement* (Brussels: European University Association).

Vettori O., Lueger M., Knassmüller M. (2007) 'Dealing with Ambivalences. Strategic Options for Nurturing a Quality Culture in Teaching and Learning'. In European University Association: *Embedding Quality Culture in Higher Education*.

A Selection of Papers from the 1st European Forum for Quality Assurance (Brussels: European University Association), 21–27.
Weaver, W., Shannon, C.E. (1963). *The Mathematical Theory of Communication*. University of Illinois Press.
Weick K. E. (1994) 'Organizational Culture as a Source of High Reliability'. In Tsoukas Haridimos (ed.) *New Thinking in Organizational Behaviour* (Butterworth-Heinemann: Oxford), 147–162.
Westerheijden D. F., Stensaker B., Joao Rosa M. (2007) 'Introduction'. In D. F. Westerheijden, B. Stensaker, M. Joao Rosa (eds) *Quality Assurance in Higher Education: Trends in Regulation, Translation and Transformation* (Dordrecht: Springer), 1–11.
Williams P. (2009) 'External Quality Assurance in the EHEA: the Role of Institutional and Programme Oriented Approaches'. In Austrian Agency for Quality Assurance (AQA) (ed.) *Trends of Quality Assurance and Quality Management in Higher Education Systems* (Vienna: Facultas), 15–23.
50+20 (2012) '50+20 Management Education for the World', http://50plus20.org/5020-agenda, date accessed 30 April 2014.

# 4
The Role of Assessment and Quality Management in Transformations towards Sustainable Development: The Nexus between Higher Education, Society and Policy

Clemens Mader

Introduction

Local as well as global challenges including financial crises, climate change, poverty, social inequalities and loss of biodiversity can only be addressed successfully through initiatives and innovations that take a long-term perspective and are developed and implemented holistically and inclusively. This requires taking into account a variety of values and perspectives, and figuring out how best to link the disciplines and create relevant and feasible scenarios for sustainable development. This, in turn, requires people with systems thinking, interpersonal, anticipatory, strategic as well as normative competences (Wiek et al., 2011). As higher education institutions (HEIs) are the incubators of innovation and the producers of future decision-makers, they have the potential to become transformational institutions for society as a whole – to address not only the needs of today but the needs of tomorrow and the development of a sustainable future. But to do this they need to transform themselves (Rio+20 Treaty on Higher Education, 2014). And this poses a major change agenda as it requires them to develop a 'fit for purpose' institutional management, shared leadership through inclusive governance and open processes and education and learning programmes that enable people to take agency and undertake innovation through transdisciplinary research.

To provide sustainability solutions and educate future decision-makers with the necessary competences, to successfully introduce transdisciplinary education and research, to work productively with their communities, to foster a more productive nexus between society, policy and their own work (Sedlacek, 2013), to reshape their operations, leadership and change management, universities need to have in place a robust, proven quality management system that responds to social, economic and environmental challenges (Mader C., 2013b).

This chapter will provide an insight into the role of quality assessment and management in transformations towards sustainable development by fostering a closer nexus between HEIs, society and policy.

Transformative higher education

The key role of education in fostering sustainable development was highlighted in Agenda 21 (UNCED, 1992) and a decade later in 2002 at the Johannesburg Summit for Sustainable Development. This was followed by the United Nations declaring the period 2005–2014 as the United Nations Decade of Education for Sustainable Development (UN DESD) (UNESCO, 2009; Sedlacek, 2013). Although a wide range of sustainability initiatives had been under way in universities prior to 2005 it was during the DESD that that the interface between higher education and society was given particular focus. A good example of this was the establishment around the world, with the support of the UN University's Institute for Advanced Studies of Sustainability (UNU-IAS), of a network of Regional Centres of Expertise on Education for Sustainable Development. The initiative was launched in 2005 with eight RCEs and by mid-2014 some 130 RCEs have been established. Each RCE supports local multi-stakeholder networks including schools, NGOs, public and private instrumentalities, businesses and higher education institutions. HEIs have a major role as knowledge generating, exchanging and transferring institutions within RCEs (Mochizuki and Fadeeva, 2008). Through participation in RCE networks, institutions have the opportunity to exchange knowledge and experience in the course of mutual learning activities and inspire socially relevant research and education by transdisciplinary activities. These transdisciplinary activities which centre around the social, cultural, economic and environmental sustainability projects most relevant to the local region, are informed by the traditional knowledge and experience of regional practitioners as well as by the research expertise and global links of the host university. It is through this quality assured network that viable locally relevant solutions to the challenges of sustainability are addressed and direct linkages to the way in which parallel challenges are being addressed elsewhere in the world are facilitated. This HEI–society nexus is proving to be a powerful way to generate, implement and share high-quality solutions and represents a significant transformation of the way in which HEIs can operate to support social transformation towards sustainable development.

The Graz Declaration on Committing Universities for Sustainable Development (UNESCO, 2005) that was concluded at the launch conference for the UN DESD in 2005 points out the distinctive responsibility universities bear for society, how they should give sustainable development fundamental status in their strategy and activities, and how they need to work reciprocally with society through transdisciplinary education and research activities. It also calls for ministries to shape their quality criteria for universities according to sustainability standards.

In 2012, universities and UNESCO took stock of what has been achieved over the previous seven years in the transformation of HEIs towards sustainable development. In the course of the Rio+20 UN Conference on Sustainable Development in Rio de Janeiro, Brazil, two key documents for HEI were produced: the Higher Education Sustainability Initiative and the Rio+20 Treaty on Higher Education.

Initiated by a group of UN partners in the run-up to the Rio+20 conference, the Higher Education Sustainability Initiative (HESI) identified activities undertaken by universities around the world to action the objectives of the DESD. A direct outcome of this work was the 'Platform for Sustainability Performance in Education' (http://www.eauc.org.uk/theplatform/home), which has now been launched by the United Nations Environmental Programme. It includes guidance on assessment processes for the area and a list of sustainability assessment tools to support HEIs in adopting strategic approaches to implement and monitor their sustainability commitments (HESI, 2014).

At the same time the Rio+20 Treaty on Higher Education was developed by the COPERNICUS Alliance, the European Network on Higher Education for Sustainable Development, the International Association of Universities (IAU) and the United Nations University (UNU) in conjunction with 31 higher education agencies, institutions, associations and student networks from around the globe (Rio+20 Treaty on Higher Education, 2014). What is significant about the Rio+20 Treaty for the topic explored in this chapter is its focus, inter alia, on transformation and quality. The treaty promotes the following eight principles:

1. To be transformative, higher education must transform itself;
2. Efforts across the higher education system must be aligned;
3. Partnership underpins progress;
4. Sustainable development is an institutional and sector-wide learning process;
5. Facilitating access to the underprivileged;
6. Inter- and transdisciplinary learning and action;
7. Redefining the notion of quality higher education;
8. Sustainable development as a whole-of-institution commitment. (Rio+20 Treaty on Higher Education, 2014).

The Treaty makes a series of recommendations on how these principles of transformation and quality assurance for embedding sustainable development in the core work of HEIs might best be put into practice. These are structured into immediate, short-term (within 3 years), medium term (within 13 years) and long-term (more than 13 years) actions. The Treaty argues that only through the active participation of each HEI in transforming its institutional practices of education, research, operations and outreach can a global transformation take place. With this in mind points 1–4 of the Treaty focus on system-wide transformation whereas points 5–8 focus mainly on local agency. In doing this the Treaty calls for a whole-of-institution commitment that quality assures an integrated focus on sustainability in campus management, curriculum, research and student and community engagement activities.

By June 2013 a total of 272 organizations had made commitments to the Higher Education Initiative and by April 2014, approximately 100 institutions had signed the HE Treaty. Engaging the disengaged remains a key challenge for this movement, with organizations like the COPERNICUS Alliance, and similar regional and global Higher Education for Sustainability organizations, promoting the agenda among hitherto uninvolved HEIs. Declarations like those discussed above and the networks that support them are important but are only the first step towards change in practice.

In order to take the next steps towards mainstreaming, the quality of what has been achieved needs to be assured, documented, scaled up and promoted. This can be facilitated through the many national and regional groups working on transforming their research, curriculum, engagement activities and operations toward great focus on the key pillars of sustainability. At an international level the UNESCO-based International Association of Universities represents more than 600 HEIs and organizations from some 120 countries. It gives focus to leadership in higher education, provides a global network and has sustainable development as one of its seven priority themes and targets. In the course of its 2014 international conference on 'Blending Higher Education and Traditional Knowledge for Sustainable Development' held in Iquitos, Peru, an outcome of the conference was the Iquitos Statement on Higher Education for Sustainable Development (International Association of Universities, 2014a), which was developed in preparation for the UNESCO World Conference on Education for Sustainable Development, taking place in Japan in November 2014. Through this statement IAU calls strongly upon organizations, institutions and policy-makers to mainstream sustainable development concepts, actively promote leadership for sustainability in higher education (Scott et al., 2012) and mobilize far more human, organizational and financial resources, to achieve sustainable development. The recently launched interactive Portal on Higher Education for Sustainable Development (International Association of Universities, 2014b) will be an important resource platform for knowledge exchange and mainstreaming sustainability in the 600 university members of IAU.

Transformation through higher education: think global act local

All of the declarations, statements and treaties identified above address five key transformation and quality principles: leadership and vision, social networks, participation, education for sustainable development, and learning and research integration. The principles accord with the Graz Model for Integrative Development (GMID) which describes the interrelation of the five principles as a process of transformation (Mader, 2012). The model can also be used for the assessment of the transformative potential of institutional and social sustainability processes (Mader, 2013).

In the following section, the five principles will be described in the context of a short analysis of five case studies: (1) leadership and vision at Heliopolis University, Egypt; (2) social network and participation at the University of Graz, Austria and RCE Kitakyushu, Japan; (3) education and learning at Africa Nazarene University, Kenya; and (4) research integration at Leuphana University Lüneburg, Germany. Throughout all of the five cases we can observe the whole-of-institution approaches that were also called for within the Rio+20 Higher Education Treaty.

(1) Universities need to create *a strong vision for sustainable development and take agency for it*. Agency needs to be taken through a *shared leadership approach* enabling the faculty, students, management and relevant stakeholders to undertake linked active leadership roles in the transformation process.

The Heliopolis University for Sustainable Development has taken leadership in a transformation of society towards sustainable development. The Sekem Development Foundation established the university in Cairo, Egypt, in 2012. With more than 30 years' history in supporting sustainable social, environmental and economic development in Egypt, the Sekem Group of companies and the Sekem Development Foundation established this new university as a linking and leveraging mechanism for transformation towards sustainable development within the Egyptian community. The university's mission is to:

> empower our students to be the champions of sustainable development in different spheres of life. We provide a place where new ideas meet fertile ground for further research and teaching. Our education combines teaching, research, and practice with a uniquely humanistic core programme, developing curious and creative minds. This will prepare the leaders of tomorrow to competently act and reflect upon the decisions awaiting the society. (Heliopolis University, 2014)

To enact this commitment to sustainable development demands active engagement and shared leadership from a variety of stakeholders, including the university's faculty, management and students, along with its business partners, NGOs and public institutions.

(2) Through active *outreach* and *involvement with all relevant stakeholders* the university can identify the real needs of society, the economy and the environment, locally and globally. In this approach HEIs have the opportunity to contribute practically to a regional transformation towards sustainable development through active stakeholder exchange, openness and inclusiveness and engagement as well as through education and relevant research initiatives for sustainable development (Mader M. et al., 2013).

One example of how to engage with society is seen at the University of Graz, Austria. The *Megaphon Uni* is a programme that promotes an open and inclusive university. The aim of the Megaphon Uni at the University of Graz is to provide education without barriers and ensure that the knowledge and socially responsible research of the university finds its way to society through easy-to-understand and practically oriented classes. Educational barriers are being reduced by targeting people who, because of their cultural or social backgrounds, have not had the opportunity or preparation to become involved in higher education. At the same time, the two-way approach of the Megaphon Uni helps ensure that higher educators become more familiar with the tasks, challenges and environments of social aid institutions in the region of Graz (University of Graz–Megaphon Uni, 2014). This also enables the university to identify regionally relevant social challenges that can be taken up in the educational and research programmes of the university. It is in this way that the Megaphon Uni is helping to contribute to the University of Graz itself as well as to the transformation of their society that surrounds it.

RCE Kitakyushu, in Japan, provides another example of how an HEI on local scale can build a close bridge to society through participation in transdisciplinary education and research programmes.

In the city of Kitakyushu, located in western Japan, the local ESD council is the main driving force for the RCE Kitakyushu. The council consists of 71 organizations and 37 individual members that include universities, research institutes, government and business sectors and many NGOS and community-based organizations. In 2013 RCE Kitakyushu, together with ten local universities, established the 'Manabito ESD station' in the city centre where the community and the local universities can interact, exchange good practice projects and ideas with each other and raise questions about the local transformation towards sustainable development. Manabito ESD station helps develop people's capabilities to build sustainable development movements and provides them with the relevant skills to act sustainably. University students get into partnerships with neighbourhoods to identify and implement common activities and have shared classes with citizens on ESD (Manabe, 2014). Through Manabito ESD station not only do universities reach out to the community but essentially the community can 'reach in' to universities to exchange information on local challenges, research demands and educational interests as well as their expectations and quality assumptions for HEI.

(3) Through *education, values and knowledge can be shared, communicated and reflected*. Education for sustainable development is a concept that has the main aim of developing a variety of competences necessary for contributing productively to the sustainable future of society, the economy and the environment (Hopkins, 2012; Lotz-Sisitka, 2014). Through *learning*, experience and knowledge are reflected upon and the interrelations between a variety of system influences can be analysed

(Lotz-Sisitka, 2014). In transformative development through *interdisciplinary, inter-cultural and inter-generational learning* actors develop an understanding of the system they are operating within (Barth and Michelsen, 2013) and the variety of values, cultures, knowledge backgrounds, environmental or economic influences that must be taken into consideration for decision-making. Universities can create these kinds of learning environments through transdisciplinary formats that build a bridge between the university and the society.

Africa Nazarene University (ANU) is located in Nairobi, Kenya. It enables students to learn with and for society. Students from the NGO *enactus*, together with their university, have initiated a project called UFO: unleashing farmers' opportunities. ANU has three core values: Character, Competence and Community. As part of the Community Engagement element of the university's values the university supports its student clubs and faculty in reaching out and serving the community. The UFO project focuses on farmers in an area of Kenya which is perpetually drought-stricken. enactus, as a club at ANU, decided to respond to this challenge by analysing ways that a community of small-scale farmers in Eastern Kenya could mitigate the impact of drought. The approach was first to exchange information about the needs and available resources, knowledge and traditions of the community. The next step was for the students to demonstrate that they could provide feasible and innovative solutions to the identified challenges. Subsequently, students trained farmers in the fields of rain-water harvesting and fruit farming using improvised drip irrigation via recycled plastic water bottles to bottlefeed the plants (Karimi, 2014).

This project at ANU demonstrates the impact that co-creative processes between higher education and society can have. Through knowledge exchange between students and farmers, students experience the farmers' challenges and can support them by developing new irrigation systems that support farming. The use of recycled plastic bottles draws on a material that, although not ideal, is both inexpensive and available. It is a simple example of how you can 'make food from waste' and how a collaboration between students and the community can contribute to the social, economic and environmental sustainability in the region. This case combines aspects of leadership, social networking, participation, education and learning as well as research. It is in this way that the university helps students engage actively in real-world learning and change for sustainability.

(4) Research integration is the fifth principle of the Graz Model and, using a *transdisciplinary approach, seeks to develop a shared ownership of*

innovation and practical outcomes from a process of mutual exchange between the HEI and society. *Transdisciplinary research* addresses the most pressing social, cultural, environmental and economic challenges facing a region in an integrated way. It uses a process of *mutual learning and innovation* that applies the knowledge of all the involved and affected stakeholders (Walter et al., 2007). This approach is consistent with two key principles in the Rio+20 Higher Education Treaty: partnerships for progress (principle 3) and inter- and transdisciplinary learning and action (principle 6).

A pioneer in running institutionalized transdisciplinary research seminars is *Leuphana University Lüneburg*, located in northern Germany, 60 km south-east of Hamburg. The university has sustainability as one of its core pillars, along with humanism and an action orientation. Through a whole-of-institution approach, the university combines research, education and institutional activities and, as such, seeks to transform itself as well as the region it serves. Its transformation towards a focus on sustainable development is embedded throughout its entire study programme. All students, no matter which programme they choose, undertake the same classes during the first semester of their studies at the university. These classes give focus to the history, methods and the social responsibility of research, with a core element being research on sustainability and the nature of transdisciplinarity. Key topic themes like the meaning of 'well-being' in 2013 or 'Climate, Change and Justice' in 2014 are carried forward through all seminars during the semester and at the end of the semester, students present the outcomes in a self-organized public conference. After this first semester students follow up with a one-year transdisciplinary research seminar. In this seminar, they develop a relevant research plan during the first semester and conduct the research during the second. A close partnership with key players and groups in the region helps ensure that socially relevant challenges are taken up in each research project. At the end of the semester conference students and their regional partners present the outcomes of their research noting how it has concretely contributed to the sustainable development of the region. Through these seminars and the close exchange between regional actors and Leuphana University Lüneburg, its students, faculty and management get key insights into the challenges of the region and to use their potential and skills to contribute to its sustainable development. This integrative approach combines strong vision and shared leadership in which regional communities and actors from the university take responsibility to jointly initiate and deliver research and educational activities. The outcomes of this approach are helping to identify relevant areas for follow-up in the university's research agenda and helping to positively influence the university's profile and role in the region.

Integrative development is key for transformation

Through this key process of integrative development for transformation via shared leadership, the development of a shared change vision, analysis of needs identified from the networks of stakeholders through their inclusive participation in education for sustainable development as well as in research activities, the university obtains a holistic understanding of how it needs to transform itself and how it can contribute to the transformation of society as a whole. These principles of integrative development (Mader, 2013) serve as guiding model for the assessment of the transformative potential (Adomssent, 2013) of each HEI (Graz Model for Integrative Development, GMID), whereas the Rio+20 Higher Education Treaty shows fields of action that universities can follow to become transformative.

To enact the above agenda it is necessary to look specifically at the role of quality assurance, tracking and improvement and the extent to which there is a nexus between higher education, policy and society in their perceptions of quality and its assessment.

Quality management and assessment to enable transformation in and by higher education institutions

Consequently it is important that the focus, tools and processes of the quality systems that underpin them (for example performance criteria, quality assessment and audit processes, quality rankings and local quality and funding systems) actively support transformation towards sustainability in our HEIs.

Figure 4.1 shows the iterative process of transformation that is enabled through assessment and quality management in one university. HEIs, as well as policy and society, can shape and are shaped by a wide range of external change forces, including changes in the society, the environment and the economy. Quality management, assessment and the scale-up of successful changes in education and research for sustainable development interact with each other. Each is discussed in the sections that follow.

Quality management

According to Schneidewind (2014) and Dyllick and Muff (2014) universities are undergoing a change of perspective from 'inside-out' towards

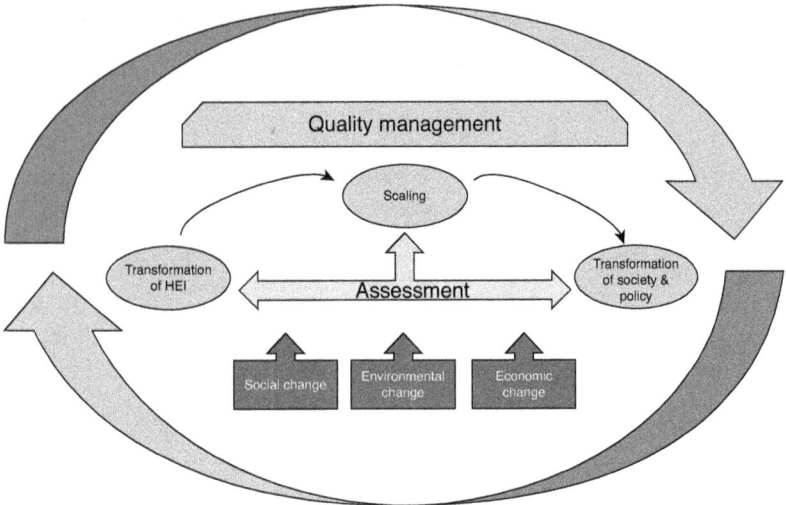

Figure 4.1 Iterative process of transformation enabled through quality management and assessment

'outside-in'. To achieve a transformation of the HEI system towards sustainable development in the context of the 21st century, a closer exchange and more co-creation of research and teaching agendas in partnership with society is, as already noted, necessary (Schneidewind, 2014). To achieve this transformation the perceptions of HEIs, policymakers and society regarding what makes a sustainable HEI need to be transformed. According to Pratasavitskaya and Stensaker (2010) the global aim of quality assurance in HEIs is to secure and develop quality of education and research. From a sustainability perspective, this focus needs to be extended to assuring the quality of the impacts which the university has on the society, the environment and the economy. As a consequence Holm et al. (2014) recommend that universities should modify their existing QM tools to embed education for sustainable development into the core of quality management in higher education.

Quality management systems in higher education need not only to be reflective of the sustainability of HEIs and inclusive of the stakeholders necessary to support their transformation (outside-in) but they also need to assist with the transformation of society and policy. Globally, some potentially relevant external quality assurance systems and standards include ISO 9001, ISO 14001 and the Environmental Management Auditing Scheme (EMAS) for environmental management along with ISO 26000, which focuses on corporate social responsibility (CSR). These are framed as total quality management systems and take responsibility for relations with stakeholders into account (Holm et al., 2014). However, although they provide a relevant report from a particular perspective, they do not take a holistic one or take the development of competences into consideration (Yarime and Tanaka, 2012). Furthermore, ISO systems are strongly focused on the process management of organizations and do not give emphasis to the quality and impact of engagement with external stakeholders, factors which are so important for real organizational and systemic transformation. Consequently they are purely focused on the inside-out process. Another reason why these systems are not suited for use in higher education is that their origin is in business and industry, not in knowledge production. In this regard Juha Kettunen in her paper 'Integration of Strategic Management and Quality Assurance' in HE, notes the absence of quality systems to assess the quality of HEI outreach and engagement within the community and region (Kettunen, 2011). The EMAS system, which has been used by Leuphana University Lüneburg in Germany for a number of years, helps address this need, however it only gives focus to engagement around the environmental aspect of sustainable development.

To be transformative, the quality management systems of HEIs need to validly assess all of the processes and systems which make up the institution's activities, in particular its objectives for sustainable development.

Assessment and quality systems towards transformation of HEIs

Institutional structures, operations, research and educational outputs are all strongly influenced by the quality criteria used and the institution's understanding of the concept of 'quality'. Quality systems provide guidelines for HEIs that have the potential to be transformative if they are developed though participative and integrative processes. This aligns with principle 7 of the Rio+20 Treaty on Higher Education, which calls for current performance indicators being directly aligned with the core missions and responsibilities of higher education (Rio+20 Treaty on Higher Education, 2014). According to Ness et al. (2007, p.499) the purpose of sustainability assessment is to 'provide decision makers with an evaluation of global to local integrated nature-society systems in short and long term perspectives in order to assist them to determine which actions should or should not be taken in an attempt to make society sustainable' (p. 499). The assessment of the unique, local context in which the HEI operates is, as noted earlier, a key factor in determining how the HEI needs to change to become transformative towards sustainable development. A variety of assessment approaches are provided by the Platform for Sustainability Performance in Education operated by the United Nations Environmental Programme (HESI, 2014).

Although universities serve such as an important role in society, they only rarely reflect and report on their impact on society (Scott et al., 2012; Adams, 2013). Whereas in industry and business it is often expected that companies will report on their sustainability actions (in some cases by law, in others by stockholders), this focus is yet to become embedded in the quality reporting systems of HEIs. Carol Adams from Monash University in Australia describes some potential reasons for this lack of sustainability assessment in higher education.

- siloed thinking and funding in both functions and disciplines, encouraged both by the way universities tend to organize into academic disciplines and by adherence to rigid or traditional norms about what constitutes an appropriate university structure and focus;
- territorialism and leadership styles which are not compatible with the collaboration required to integrate and embed sustainability (Adams et al., 2011, Scott et al., 2012);
- mandatory reporting requirements for universities which often focus on trivial issues rather than the bigger picture or on material impacts and which have, in many countries, not yet been adapted to incorporate sustainability issues;
- a lack of focus on the business case (such as: concerns of future students and those that will employ them; cost savings through the adoption of carbon management and energy reduction plans; increased staff satisfaction and the ability to attract good staff, particularly those in non-academic roles);
- limited push from sector associations, trade unions and student groups; and
- little conception of what a best practice university in sustainability might look like. (Adams, 2013, p. 386)

These reasons for a lack of quality assessment of sustainability in our universities indicate an absence of the transformative potential principles described with GMID (shared leadership, social inclusion, participation, education and learning for sustainable development, transdisciplinary research). Governance structures within HEIs are often strongly hierarchical and mitigate against players at different levels taking shared responsibility for cross-university action on sustainable development. HEIs require more inclusive leadership forms, where opportunities for agency towards sustainable development are provided for staff from all levels, along with students and key players from the region. Another important aspect raised by Adams is the lack of incentives and skills to work in an inter- and transdisciplinary way. To transform the HEI system, society and policy, HEIs need to exchange with other institutions, learn from one another and support each another through common strategies. An example would be regional or national networks of HEIs that focus on the institutional and system-wide transformation of higher education and the society. Together institutions can agree upon new and sustainability oriented set of quality criteria for the HEI system. Stakeholders, global trends and locally relevant challenges can be analysed and reflected in common activities.

Scaling towards systemic transformation in higher education, society and policy

Scaling starts with leadership: someone who brings together the right team, helps them shape and implement the desired sustainability initiative and makes sure that, should they leave, there are others who can sustain what has been achieved. Key players can be both internal to a university and external from the community its serves. Also, institutionalizing the initiative within the system can help to build a firm foundation for scaling up a successful sustainability innovation. Louis C. Boorstin, former deputy director at the Bill and Melinda Gates Foundation, reports from his experience:

> Yet such innovations can succeed in the long run only if they are embedded in the local systems. Social innovators who seek scaled impact, should focus on altering how pivotal institutions set policies, allocate funding and deliver services on the ground. Applying influence at institutional leverage points can generate long-term, wide-scale improvements. (Boorstin, 2013, p. 14)

The same author describes later in his paper that, to scale innovations, ongoing assessment is most important for the sustained implementation of a desired change and for effective scale-up. However, in many cases assessors and implementers work separately. But, if they work in teams and/or exchange is fostered, the quality of assessment improves and implementation of assessment outcomes finds much more relevance (Boorstin, 2013).

Assessment (both to prove and improve quality and foster ownership) is, therefore, the basis for making scaling possible. Through assessment policy and practice responses that are appropriate to scale the assessed initiative are identified (Dovers, 1995). Valid assessment can foster transformation and lead to the generation of system understanding and its relations. This can be transferred to other contexts and scales. Through participation of stakeholders from various roles and backgrounds a holistic understanding of the system can be generated. This is necessary to scale initiatives out to other regions or scale-up for systemic implementation. It is through exchange and co-creation between assessors and implementers that the adaptation of practices for new system environments takes place.

This analysis of the role of assessment in scaling demonstrates how important the declarations, treaties, networks and exchange opportunities described earlier are – as they are the mechanism through which assessed successes in transformation can be taken up and adapted in other locations. At the same time, once transformation of society and policy takes place, a rethinking of quality systems and management takes place. This can already be observed in countries like Austria where criteria related to the assessment of sustainability engagement and effectiveness are reflected in the performance agreements between the Ministry of Science, Research and Economy and the universities. At a global level the United Nations General Assembly has stated in the Rio+20 outcome document 'The Future We Want' (United Nations General Assembly, 2012) that the nexus between science/society and science/policy (Art. 48) should be strengthened. Examples like this and others outlined in the paper show that first fruits of institutional and social transformation can be harvested. As demonstrated in Figure 4.1, the process entails an ongoing, upward spiral of change, driven by interrelated processes and factors in which effects in one of the areas imply impacts of transformation in the others.

Conclusions

In this chapter, the role of quality assessment and quality systems in transformation towards sustainable development through the nexus between higher education, society and policy has been studied.

HEIs are attracting more and more public focus on the quality of their education and research (De Villiers et al., 2014). Global initiatives calling for a strong transformation towards sustainable development also highlight the important role of quality assessment, improvement and management.

The chapter has outlined the framework of the iterative process that is required to transform HEIs and foster a closer nexus between the HEI system, society and sustainable development policy, quality management and reporting. The initiatives from around the world reviewed in the chapter provide a wide range of proven case studies of how transformative development for a sustainable future can be implemented. The meaning of transformative development has been modelled through the Graz Model for Integrative Development that brings together five key change principles: leadership and vision, social network, participation, education and learning and research integration. How these principles can be applied has been illustrated in the case studies and how they can contribute to quality assessment and system transformation has been suggested. In Figure 4.1 these quality management and assessment methodologies and principles for transformation are seen to stand at the core of the framework as the key bridges for transformation.

References

Adams C. A. (2013) 'Sustainability Reporting and Performance Management in Universities – Challenges and Benefits'. *Sustainability Accounting Management and Policy Journal*, 4 (3) 384–392.

Adams C. A., Heijltjes M. G., Jack G., Marjorobanks T., Powell M. (2011) 'The Development of Leaders Able to Respond to Climate Change and Sustainability Challenges: the Role of Business Schools'. *Sustainability Accounting, Management and Policy Journal*, 2 (1) 165–171.

Adomssent M. (2013) 'Exploring Universities' Transformative Potential for Sustainability – Bound Learning in Changing Landscapes of Knowledge Communication'. *Journal of Cleaner Production*, 49, 11–24.

Barth M., Michelsen G. (2013) 'Learning for Change: An Educational Contribution to Sustainability Science'. *Sustainability Science*, 8 (1) 103–119.

Boorstin L. C. (2013) 'The Quest for Scale'. *Stanford Social Innovation Review*, 11 (4) 13–14.

De Villiers C., Chen S., Jin C., Zhu Y. (2014) 'Carbon Sequestered in the Trees on a University Campus: A Case Study'. *Sustainability Accounting Management and Policy Journal*, 5 (2) 149–171.

Dovers S. R. (1995) 'A Framework for Scaling and Framing Policy Problems in Sustainability'. *Ecological Economics*, 12 (2) 93–106.

Dyllick T., Muff K. (2014) 'The Business Sustainability Typology. A Briefing for Organizational Leaders and Academic Scholars', working paper, March 2014, http://papers.ssrn.com/sol3/papers.cfm?abstract_id=2368735, date accessed 29 April 2014.

Heliopolis University (2014) 'Vision and Mission', http://www.hu.edu.eg/HUvision, date accessed 20 April 2014.

HESI (2014) 'Higher Education Sustainability Initiative', http://sustainabledevelopment.un.org/index.php?menu=1073, date accessed 24 April 2014.

Holm T., Sammalisto K., Vuorisalo T. (2014) 'Education for Sustainable Development and Quality Assurance in Universities in China and the Nordic Countries: a Comparative Study'. *Journal of Cleaner Production*, DOI: 10.1016/j.jclepro.2014.01.074.

Hopkins C. (2012) 'Reflections on 20+ Years of ESD'. *Journal of Education for Sustainable Development*, 6 (1) 21–36.
International Association of Universities (2014a) 'Iquitos Statement on Higher Education for Sustainable Development Draft 2 for circulation', http://www.iitk.ac.in/infocell/announce/convention/papers/Strategy%20Learning-02-Juha%20Kettunen.pdf, date accessed 25 April 2014.
International Association of Universities (2014b) 'Portal on Higher Education for Sustainable Development', http://www.iau-hesd.net/en, date accessed 28 April 2014,
Karimi R. (2014) Nazarene University, Nairobi, Kenya, email correspondence 18 April 2014.
Kettunen J. (2005) 'Integration of Strategic Management and Quality Assurance', http://www.iitk.ac.in/infocell/announce/convention/papers/Strategy%20 Learning-02-Juha%20Kettunen.pdf, date accessed 28 April 2014.
Lotz-Sisitka H. (2014) 'Radically Reshaping Higher Education for the Future', SciDev.Net, http://www.scidev.net/global/education/opinion/radically-reshaping-higher-education-for-the-future.html, date accessed 21 April 2014.
Mader C. (2012) 'How to Assess Transformative Performance towards Sustainable Development in Higher Education Institutions'. *Journal of Education for Sustainable Development*, 6 (1) 79–89.
Mader C. (2013) 'Sustainability Process Assessment on Transformative Potentials: the Graz Model for Integrative Development'. *Journal of Cleaner Production*, 49, 54–63.
Mader C., Scott G., Razak D. A. (2013b) 'Effective Change Management, Governance and Policy for Sustainability Transformation in Higher Education'. *Sustainability Accounting, Management and Policy Journal*, 4 (3) 264–284.
Mader M., Mader C., Zimmermann F. M., Görsdorf-Lechevin E., Diethart M. (2013) 'Monitoring Networking between Higher Education Institutions and Regional Actors'. *Journal of Cleaner Production*, 49, 105–113.
Manabe K. (2014) 'Expansion of Cross-Sectoral and Practical ESD Education "Kitakyushu Manabito ESD Station"'. In Y. Bessho, *Proceeding of International Symposium – ESD and Universities 2, Searching for Sustainable Local Communities* (Nagoya: Nagoya City University), 17–23.
Mochizuki Y., Fadeeva Z. (2008) 'Regional Centres of Expertise on Education for Sustainable Development (RCEs): An Overview'. *International Journal of Sustainability in Higher Education*, 9 (4) 369–381.
Ness B., Urbel-Piirsalu E., Anderberg S., Olsson L. (2007) 'Categorising Tools for Sustainability Assessment'. *Ecological Economics*, 60 (3) 498–508.
Pratasavitskaya H., Stensaker B. (2010) 'Quality Management in Higher Education: Towards a Better Understanding of an Emerging Field'. *Quality in Higher Education*, 16 (1) 37–50.
Rio+20 Treaty on Higher Education (2014) 'Rio+20 Treaty on Higher Education', http://hetreatyrio20.com/, date accessed 20 April 2014.
Schneidewind U. (2014) 'Von der nachhaltigen zur transformativen Hochschule. Perspektiven einer "True University Sustainability"'. *umweltwirtschaftsforum*, DOI: 10.1007/s00550-014-0314-7.
Scott G., Tilbury D., Sharp L., Deane E. (2012) *Turnaround Leadership for Sustainability in Higher Education Institutions* (Sydney: Office for Learning and Teaching, Department of Industry Innovation, Science, Research and Tertiary Education).

Sedlacek S. (2013) 'The Role of Universities in Fostering Sustainable Development at the Regional Level'. *Journal of Cleaner Production*, 48, 74–84.
UNCED (1992) 'Agenda 21', http://www.un.org/esa/dsd/agenda21/, date accessed 20 April 2014.
UNECE (2012) *Learning for the Future – Competences in Education for Sustainable Development* (Geneva: United Nations Economic Council for Europe).
UNESCO (2005) 'Graz Declaration on Committing Universities to Sustainable Development', http://www-classic.uni-graz.at/geo2www/Graz_Declaration.pdf, date accessed 24 April 2014.
UNESCO (2009) 'Education for Sustainable Development and Climate Change', Policy Dialogue 4 (Paris: UNESCO).
United Nations General Assembly (2012) 'The Future We Want', A/res/66/288 (New York: UN).
University of Graz–Megaphon Uni (2014) 'University of Graz, Megaphon Uni', http://megaphonuni.uni-graz.at/de/ueber-die-megaphonuni/, date accessed 20 April 2014.
Walter A. I., Helgenberger S., Wiek A., Scholz R. W. (2007) 'Measuring Societal Effects of Transdisciplinary Research Projects: Design and Application of an Evaluation Method'. *Evaluation and Program Planning*, 30 (4) 325–338.
Wiek A., Withycombe L., Redman C. L. (2011) 'Key Competencies in Sustainability: A Reference Framework for Academic Program Development'. *Sustainability Science*, 6 (2) 203–218.
Yarime M., Tanaka Y. (2012) 'The Issues and Methodologies in Sustainability Assessment Tools for Higher Education Institutions: A Review of Recent Trends and Future Challenges'. *Journal of Education for Sustainable Development*, 6 (1) 63–77.

Part II

The Meaning and the Role of the Internal Quality Assurance and Its Interplay with External Quality Approaches in Supporting HE Sustainability Transformation

5
Drivers for Change in the Austrian University Sector: Implications for Quality Management
Nadine Shovakar and Andrea Bernhard

Introduction

Education is broadly seen as a decisive factor to *'improve productivity, employability, nutrition and healthcare, and general prosperity'* (OECD, 2011, p. 54), and this has led to an increasing interest by governments in education policy. In the so-called knowledge-based economies, this interest has placed higher education at the centre of national competitiveness agendas (Sursock and Smidt, 2010, p. 14). For example, the EU bases its 2020 strategy on 'smart growth through more effective investments in education, research and innovation' (European Commission, 2014). As a consequence of both internal and external change pressures over the past decade, Europe has undergone many ground-breaking reforms. Managing these change processes – especially if initiated by external players – takes time and conviction, and the question arises as to whether universities have managed to stay well positioned in these turbulent times and if so which strategies have proven to be successful?

According to the Oxford Dictionary, change management is 'the management of change and development within a business or similar organization' (Oxford Dictionaries, 2014). The European University Association (EUA) states in its Trends Report VI (Sursock and Smidt, 2010, p. 91) that, after ten years of change related to the Bologna Process, *'Successful implementation of Bologna is partly conditional on the capacity of institutional leaders to bring institutional coherence to a multi-dimensional change agenda, and to explain, persuade and motivate staff members, and students.'* Thus, it is not only about drafting strategies but also about convincing the people involved to engage with the changes the strategy foresees. According to Skordoulis (2012, p. 3), high-level executives nevertheless still underestimate the importance of involving the personnel in change processes:

> The focus in many change initiatives, as well as in the literature, is primarily on creating and designing optimal solutions for innovative ways of doing business. The main emphasis lies on pioneering and designing the change process, whereas actual implementation of solutions in practice is often considered as a mechanistic task of executing the plans.

Similarly, the quest to define a university profile and strategic priorities has to be supported by all university members. Change does not always rest comfortably with university teachers owing to their commitment to academic freedom and disciplinary loyalties, which do not automatically coincide with the university management's transformative ideas. To address this dilemma, Hans Weiler (2006, p. 44) recommends a mixed approach of being patient enough to convince people on an individual level and on the other hand to provide incentives to follow the institutional policies. Skordoulis (2012, p. 16) recommends a combination of top-down and bottom-up approaches in order to '*compete in a highly unstable and ever-changing higher education arena*' for the following reasons:

- Top-down restructuring is a management-driven process that is not necessarily consensus-seeking. In the particular culture, such a process would alienate employees further if not accompanied by a bottom-up process that empowers individuals, who need to feel like stakeholders with something to gain as well as to lose.
- Holistic viewing of a system can reveal structural changes that extend beyond localized levels of transactions.
- Many staff feel overworked and under-appreciated due to overly bureaucratic and irrelevant policies or processes.
- Both top-down and bottom-up perspectives are needed – reflecting synthetic and analytic mechanisms – in order to inform culture and process transformation.

An important precondition to support change processes in universities is the institution's governance structure which in the case of Austria has been modernized by a university reform in 2002.

The Austrian case

The Austrian higher education sector caters for roughly 360,000 students that represent approximately 4.2 per cent of the country's population. By far the largest group are being educated at the 21 public universities (81%); the rest is spread over 12 private universities (2%), 21 polytechnics (11%), 14 teacher training colleges (4%) and one university for continuing education (2%). At the public universities every fifth student has an international passport: that is twice as high as the percentage of non-Austrians of the general population (Statistik Austria, 2013, pp. 9–32).

Historically a number of significant reforms in higher education have been implemented. Universities now have a greater autonomy and under the new Universities Act (UG, 2002)[1] universities have been 'opened up' to society and economy, competition among universities has been encouraged, greater efficiencies sought and clear university profiles developed while, at the same time, this should not entail a loss of critical reflection, innovation and basic research. According to this act, public universities:

- are legal entities with *autonomous status* and have autonomy to determine their internal organization (Article 5, UG, 2002);
- are *employers* under private law (Article 108, UG, 2002);
- have full responsibility for the design of their *curricula* (Article 54 (1), UG, 2002);
- receive public funding: *three-year lump-sum* budgets (Article 12 (6), UG, 2002);
- are governed by a *rectorate*, a senate and a university council (Article 20 (1), UG, 2002).

The trade-off for autonomy is transparency as the government and the taxpayers want to know how universities spend the money they have given them. This has materialized in the different reporting procedures to the Ministry for Science and Research (for example annual financial statements and intellectual capital reports) and in the introduction of quality management systems at universities (Article 14, UG, 2002). Furthermore, the new Higher Education Quality Assurance Act (HS-QSG) came into force in 2011 and all universities must undergo an external audit by an external agency every seven years. This process has been developed parallel to the quality initiatives on a European level which were triggered by the Bologna Process.

According to an analysis of the Austrian higher education sector, which was undertaken in order to develop a master plan for its future organization, the amount of reporting required by public authorities was seen to involve an unreasonable administrative burden for universities (Loprieno et al., 2011, p. 45). Furthermore, the universities feel that the Ministry for Science and Research has so far not been making effective use of the various steering instruments it has at its disposal. In other words, the Ministry for Science and Research demands very detailed information on organizational matters in the performance agreements but fails to take decisions on a strategic level (APA, 2012). To sum it up, the Universities Act was a big step forward in increasing the autonomy of universities. However, it also has a number of limitations:

- lack of university control over their student admissions (Article 64, UG, 2002);
- lack of control over tuition fees (Article 92, UG, 2002);
- no ownership of university buildings (in most cases).

In 2011, the Austrian government put in place a national strategy on research, technology and innovation (FTI strategy), which is currently being implemented. Among other measures, the FTI strategy calls for a clustering of research fields, a sustainable model for university funding and a stronger cooperation between universities and the private sector. Universities Austria claims stronger commitment to foster basic research, which is seen as a key element in the innovation process.

Interestingly, Elmar Pichl, a high-level representative of the Austrian Federal Ministry for Science and Research, considers a bottom-up approach as the main impetus for development and innovation at public universities. Moreover, Pichl defines polytechnics and private universities as mainly driven by a top-down approach and the government-run teacher training colleges should be run by a bottom-up and top-down management mix (Pichl, 2012, p. 203).

Austrian universities in the European Higher Education Area

Austrian universities are well connected beyond the national borders, and higher education strategies are often decided at a European rather than a national level. The European Union's Modernization Agenda, the Europe 2020 strategy, the Lifelong Learning Programme, just to name a few, are influencing the universities' realities by either financial incentives or political pressure as can be seen by two examples.

On the one hand, these strategies materialize in EU research programmes such as its Seventh Framework Programme (FP7), which follow the EU's priorities and thereby also influence the universities' research agendas. Austria is an active participant in the FP7 with 2.5 per cent of the projects, or in other words the tenth position of the EU 27 according to the national monitoring project PROVISO (2013, pp. 14–55). Compared with the number of researchers per nation (2.2%) this indicates an above-average importance of the programme in the Austrian context. The Bologna Process on the other hand – at least for Austria – does not come with fresh money but with national legislation and recommendations (soft law). Yet, it can be seen as one of the biggest educational reforms in Europe, since significant changes in the areas of student-centred learning, quality assurance, widening access and so on[2] have been initiated in its name. After ten years, the European Higher Education Area (EHEA), establishment of which was a key goal of the Bologna Process, has officially been launched during the Budapest-Vienna Ministerial Conference in March 2010. Universities Austria calls for a phase of consolidation, which involves all stakeholders much more and provides suitable ways of communication and information on a national and a European level.

As can be seen by this introduction, the system in which universities in Austria are embedded is complex with many interdependent actors and stakeholders at both the national and international level. Moreover, universities themselves have structures which are characterized by a high degree of internal autonomy and democratic decision processes.

In this chapter, we will analyse how change has taken place over the past decade by analysing two cross-cutting change/transformation processes, namely the Bologna Process and Education for Sustainable Development (ESD). By reporting on the different endeavours that were initiated by the players of the Austrian higher education sector, we will lastly raise the complex question of quality and how quality management as a steering instrument in autonomous universities has helped to give transformative meaning to the policies.

Method of analysis and structure

The basis for the choice of an appropriate methodological approach and an adequate research design is the underlying research question: *how change processes take place at universities in Austria*. This article shall apply a methodological approach to deal with the change processes considering the implications of quality management tools. First of all the unit of analysis and, second, the different research instruments will be described.

The units of analysis are the 21 universities in Austria and associated stakeholders such as the Universities Austria, the Austrian Federal

Ministry for Science and Research and the European Union. There are different ways in which to look at a university sector: at system/national (national university sector), institutional or programme level but also at international level (national system embedded in an international higher education area). In this chapter all three levels are the focus. From the Austrian university sector, two selected change/transformation processes are drawn, which have had a high impact not only on Austrian higher education but also on the whole European Higher Education Area: the Bologna Process and ESD. Furthermore, the focus incorporates a study of the interplay of internal and external quality management tools to achieve a certain change/transformation process.

The research instruments of this analytical study are documentary and multivariate analysis. This chapter primarily focuses on a casestudy approach using various sources of evidence, for example surveys, documents and archival records. The data are based on secondary literature research: professional literature, a review of national legislation and policy, content analysis of professional discourse in acknowledged journals and conferences in the field of higher education, as well as the Bologna Process and ESD in particular, along with some statistical data on student numbers and higher education budget from the national database (see section 'The Austrian case' in this chapter).

The chapter shall provide an explorative but selectively systematic overview of the reviewed 21 universities by considering the usage of the ongoing transformation strategies and changes in the respective university. Thus, the focus is on a systematic analysis of the impact of two change/transformation processes – the Bologna Process and ESD – on the 21 universities along three dimensions:

1. Historical/international dimension
2. National dimension
3. Institutional dimension (performance agreements 2013–2015)

The *historical/international dimension* will describe the historical background of the respective change/transformation process by considering the international pushing factors of how these processes have started to affect the national university sector.

The *national dimension* will highlight further initiatives and activities in the field of Bologna and ESD to show the contributions of other stakeholders. Universities are crucial players in most of the initiatives; however, the impetus for change derives from external players operating at national level.

The *institutional dimension* will be the core of the analysis and highlights the future activities of the individual universities concerning the selected change/transformation processes. The analysis is based on each institution's performance agreement with the government. Every three years the Ministry of Science and Research starts a negotiation process with universities with the goal of concluding performance agreements (Article 13, UG, 2002; Bundeskanzleramt-RIS, 2002). In these contracts, the university's core business fields are defined and specific projects in these fields are agreed upon and financed. There are four areas in which universities operate according to the law:

1. Strategic objectives, academic profiles, and university and human resource development (Article 13 (1a), UG, 2002)
2. Research and the advancement and appreciation of the arts (Article 13 (1b), UG, 2002)
3. Study programmes and further education (Article 13 (1c), UG, 2002)
4. Further areas such as social goals, mobility, cooperation and further university specific goals (Article 13 (1d–i), UG, 2002).

So far there have been three performance agreements with the Ministry for Science and Research starting in 2007–2009 after a preparatory phase with a fixed allocation of the universities block grants (2004–2006). This was to ensure that the universities have the time necessary to develop their drafts for the performance agreements. The most recent negotiation process was completed in December 2012 and its results have been studied to identify the current state of projects around the Bologna Process and concerning sustainability issues. In a first step, the 21 contracts[3] will be screened for both words 'Bologna' and 'nachhaltig'.[4] Wherever a longitudinal comparison seems appropriate, these findings will be complemented by reference to previous performance agreements. We will then take a deeper look into specific sections of each report, for sustainability in the section on 'social goals' and for Bologna in the section on 'study programmes and further education'. Additionally, the field of 'quality assurance' with the primary focus on the section 'strategic objectives, academic profiles, and university and human resource development' will be screened. This additional focus on the evolution of internal as well as external quality assurance procedures reflects the third mission of higher education institutions as well as the implementation of the Bologna Process.

After this analysis of the two change/transformation processes along the three dimensions described above, a meta-analysis of the two change processes will be conducted. The internal and external quality management tools are considered as main drivers for change within the Bologna Process and ESD. Along the historical dimension, the prevailing situation and the future developments concerning the two selected change processes, the final analysis shall provide concluding remarks to the leading research question on *how change processes take place at universities in Austria.*

Transformation 1: Bologna Process

Historical/international dimension

The discussion on quality in higher education is tightly linked to the Bologna Declaration of the European Union Ministers of Education in June 1999 (European Ministers Responsible for Higher Education) and its links to the growing need for more visibility, transparency and comparability. The Bologna Process created a completely new operating context for HEIs and can be seen as a starting point for a lot of transformations throughout Europe. The initiative influenced national policies and led to the establishment of a structure and mechanism for quality assurance at the European level (Eaton, 2001, p. 41). One of the main goals of the Bologna Declaration was to create an international legal framework for the establishment of a *European Higher Education Area* (EHEA) in order to promote European cooperation in quality assurance, stronger linkages between quality assurance and recognition and foster closer cooperation between actors in these two fields at institutional, national and European levels.

Since the meeting in Bologna, every other year the Ministers of Education meet to measure the progress and set new priorities for action in their communiqués. A series of ministerial conferences from Bologna in 1999 to Bucharest 2012 (EHEA, 2012) have led to a range of important changes in the whole higher education sector, such as the formation of a European and national qualification framework, a new degree structure, the promotion of student-centred learning, a learning outcomes orientation and widened access to higher education with a special focus on the social as well as the global dimension and employability. All of these changes have had a strong impact on quality assurance issues.

According to the original time frame the EHEA should have been put into practice within ten years but this ambitious European project (mainly the implementation of a three-cycle degree structure) needed more time. The roadmap until 2020 seeks to improve and deepen the implementation of the action lines started in the first decade. The implementation of the Bologna Process and the decision of the Austrian government to take part in it took place – more or less at the same time – when Austrian public universities gained their new legal status as autonomous entities. According to Staudacher (2012, p. 77) 'Bologna' was and is the external trigger for a reform which was long overdue in the Austrian education system. However, at the beginning the implementation through legally binding norms (hard law) was prioritized. The amendment of the University Study Act in 1999 and the Universities Act 2002 established the legal basis necessary for bachelor's and master's study programmes, ECTS, the diploma supplement, joint degree programmes and doctorate/PhD programmes to be delivered. With the University Law Amendment Act in 2009 further regulations came into force (Ministry of Science and Research, 2012, p. 10):

- the possibility for study programmes in medicine and teacher education to change to the Bologna study architecture after having been initially excluded;
- all new study programmes must be structured along the Bologna system;
- increased flexibility in the duration of study of the bachelor's study programmes;
- clear qualification profiles;
- the organization of curricula to promote mobility.

The implementation of the Bologna Process primarily took place at institutional level at the universities (polytechnics and teacher training colleges respectively). Owing to the above-mentioned autonomous status universities are not bound to any directives by the Ministry for Science and Research. However, the instrument of the performance agreements[5] provides the possibility to implement measures and objectives in a concerted and coordinated way in the autonomous university sector. Already within the first performance agreement period, 2007–2009, goals and plans have been defined according to the action lines of the Bologna Process together with the universities. Within the next performance agreement period, 2010–2012, the Ministry and the universities agreed on targeted projects and objectives to implement the Bologna Process, such as the Bologna study architecture, qualification profiles, aspects of the social dimension, lifelong learning, blended learning, employability, part-time study programmes, quality assurance and mobility (Ministry of Science and Research, 2012, p. 10). The next performance agreement period, 2013–2015, will be described in detail in the subsection 'institutional dimension'. The yearly reports on the status of implementation of the Bologna action lines are published on the website of the Ministry for Science and Research.[6] These reports primarily describe additional impacts of the Bologna Process in Austria.

After the first decade of the Bologna Process, the collected experiences showed that the different university types (scientific, medical, technical, artistic) do not have the same understanding and also differ in their ways of implementing the three-cycle structure. The paradigm shift from the input to the output orientation (learning outcomes) and its consequences for teaching and learning are coming slowly. Nonetheless, the universities have already started to revise their curricula again and there are several challenges for the upcoming decade because of the implementation of the academic programmes in the field of teacher education and medicine along the Bologna structure, the increasing of the consciousness concerning the objectives of the process as well as the critical reflection on it (Westphal, 2010, p. 28).

National dimension

Universities Austria (uniko)[7] opened up a platform for intensive exchange between the vice-rectors for teaching to coordinate the implementation process and showcase examples of best practice. Especially after 2005 increasingly more activities in the field of the Bologna Process took place initiated by the universities themselves and the uniko. On the one hand universities established units responsible for Bologna-specific themes and questions ('Bologna Points') to assist their institutions internally, while on the other hand uniko tried to support universities in selected fields and worked out common measures and guidelines. The thematic fields varied from bachelor's and doctorate studies to the National Qualification Framework (NQF) (Westphal, 2010, p. 28). These activities have been strengthened in 2008 with the help of three projects related to the Bologna Process by uniko.[8]

Currently, the Austrian universities are discussing possibilities of quality improvement of the third cycle. Already in 2007 the uniko adopted 'Recommendations by Universities Austria on New-Style Doctoral Studies' to support the reforms of doctoral study programmes which have been published in 2008 (Universities Austria, 2008). Another important initiative concerning the third cycle is the ARDE project (Accountable Research Environments for Doctoral Education) supported by the European Union which has recently published its final report (Byrne et al., 2013). The European University Association (EUA), the University College Cork (UCC), the Conference of Rectors of Academic Schools in Poland (CRASP) as well as the uniko are in the project team of

ARDE. The main objective of this project was an analysis of 100 European universities with the focus on their doctoral study programmes: organization, good practice and aspects of quality assurance and improvement. Within this project an EU-wide survey on doctorate study programmes has been conducted which revealed that half of all institutions are planning reforms within their doctorate study programmes.

Furthermore, the national Bologna Follow-up Group (BFUG) coordinated by the Ministry for Science and Research has accompanied the Bologna Process in the past and is still a leading organization. In this group different higher education providers and other stakeholders (ministries, quality assurance agencies, students union, etc.) are meeting regularly to exchange the developments at national and European level. Representatives of several universities and the uniko are members of the BFUG. Currently, representatives of universities are also nominated experts within the 'Ad hoc Working Group on the Third Cycle' as well as the 'Recognition of Prior Learning Network' which are operating at European level.

Gottfried Bacher (2013)[9] reported the latest perspectives from the European BFUG and mentioned the new working plan: consolidation and combination (especially in administration); meta-goals quality, employability, mobility; reduction to four working groups (WG Reporting, WG Structure, WG Social Dimension, WG Mobility and Internationalization). According to Bacher (2013, p. 9) the Bologna goals have been implemented to the following degrees in Austria:

- the merging of the education sectors is not implemented yet;
- learning outcomes, student-centred learning, transversal competences, recognition of prior learning, lifelong learning and learning of languages are at the beginning of implementation;
- qualification frameworks, ECTS and recognition are fields which are nearly implemented;
- the bachelor's/master's structure and quality assurance are fully implemented.

Bacher's personal impression is that 'Bologna does not only bring joy but also frustration', though refers to an apt quotation from Samuel Beckett: 'Ever tried. Ever failed. No matter. Try Again. Fail again. Fail better.'

Institutional dimension (performance agreements 2013–2015)

In the following, an analysis of the performance agreements 2013–2015 of the 21 public universities in Austria will be presented. The focus is laid on the currently most important aspects derived from the Bologna Process, which will be highlighted by the universities within the next three years.

In the latest performance agreements, 13 universities have mentioned 'Bologna' or 'Bologna Process' in their performance agreements. The main context Bologna was used in can be summarized as an improvement of the already achieved Bologna implementation: adaption or continuous further development of the curricula, continuation of the Bologna Process, consistent implementation of the Bologna criteria, optimizing of the Bologna three-cycle structure, second phase of the Bologna reform, 'Bologna 2.0', Bologna revisited – quality assurance in teaching, finalization of the Bologna requirements, improvement of the Bologna implementation. Nonetheless, the performance agreements of eight universities did not refer to the word 'Bologna' at all. Out of these, at least six universities claim parts of aspects relating to the Bologna Process, for example learning outcome orientation, allocation of ECTS, implementation of bachelor's and master's study programmes for teacher education, student-centred learning, and qualification profiles/employability. In the following, selected topics concerning the Bologna Process will be highlighted in more detail.

The allocation and attribution of ECTS credits will be optimized, evaluated and/or reformed by 13 universities. Out of these, five universities focus on the 'ability of studying' and three universities plan to optimize the allocation of ECTS credits in line with the principles of the ECTS guideline of the General Directorate Education and Culture of the European Commission (2009). Two universities (University of Natural Resources and Life Sciences, Medical University of Graz) want to apply for the diploma supplement label and one university (University of Innsbruck) for the ECTS label of the European Union in 2013. Moreover, one university wants to update the ECTS label (Medical University of Graz) and one university wants to update the diploma supplement label (University of Innsbruck).

Student-centred teaching and learning is stated by seven universities as a goal for the next three years in different contexts: peer-to-peer teaching, mentoring programmes, curricula development (including the Dublin Descriptors) with compliance of the qualification profiles and the learning outcomes of study programmes, curricula reforms under consideration of the feedback of students and alumni, development of student-centred learning and self-directed learning with the support of blended learning. Ten universities have mentioned learning outcomes or outcome/competence orientation in their performance agreements as important aspects in their curricula reforms. Especially important is the assessment of single courses as well as whole models and their compliance with the defined learning outcomes.

Teacher education was another aspect linked to the Bologna Process within the performance agreements with the Federal Ministry for Science and Research. The government is currently negotiating a reform of teacher education in Austria. Until now all study programmes of teacher education at universities are still in the old structure of diploma study programmes but will be changed towards the Bologna structure of bachelor's and master's programmes. While some of the art universities want to keep the old structure, other universities have established or are in the process of establishing special units (Schools of Education, Centres for Teacher Education) which will support the curricula reform along the Bologna study structure within the next academic years.

The focus on quality as well as quality assurance within teaching/academic programmes in general and even with reference to the Bologna Process was more or less mentioned by all universities. Course evaluations are already made by all universities but initiatives for further development or extensions of course evaluations are foreseen for this period. Furthermore an external audit procedure of the internal quality management including the field of teaching/academic programmes is regulated by law. Hence, twelve universities will undergo an audit procedure within 2013–2015, three universities within 2016–2018. Two universities underwent an audit in 2012, one university will undergo a European Quality Improvement System (EQUIS) accreditation in 2013 and three universities have not yet decided when to undergo the audit procedure. In the context of alumni surveys or monitoring as well as tracking studies or labour market monitoring, 16 universities mentioned initiatives within the next three years.

Looking at these plans and goals for the next three years, the Bologna Process as the main trigger for reform has taken a backseat compared to the previous decade. Of course, most of the above-mentioned initiatives are linked to the Bologna Process but the universities are generally referring less to it and are instead naming quality assurance, quality criteria or quality assessment as triggers for change. The tenor is more towards optimization, continuation, further development, revision or improvement while the Bologna Process is more or less seen as completed.

Transformation 2: Education for Sustainable Development (ESD)

Historical/international background

The concept of Education for Sustainable Development (ESD) and thereby the link between education and sustainable development was established during the UN Conference on Environment and Development (UNCED) which was held in Rio de Janeiro in 1992. During this conference, Agenda 21 (an action plan for sustainable development) was adopted which stated in its Article 36 that education is 'critical for promoting sustainable development and improving the capacity of the people to address environment and development issues'. Moreover, the United Nations General Assembly proclaimed a UN Decade of Education for Sustainable Development (DESD) between 2005 and 2014 to further foster ESD. It designated UNESCO as the lead agency to promote and implement the Decade.

In Europe, the European University Association (EUA) took the lead in 1993 with an initiative that acknowledged the importance of higher education in ESD. The EUA invited its member universities to support and endorse a charter on ESD, which outlined ten 'change pathways' and has been ratified by 326 European universities today. In 2005, the Bergen Communiqué (EHEA, 2005, p. 4) which is part of the Bologna Process linked higher education policy with ESD stating: 'Our contribution to achieving education for all should be based on the principle of sustainable development and be in accordance with the on-going international work on developing guidelines for quality provision of cross-border higher education.'

In 2007, ten European universities founded the COPERNICUS Alliance,[10] a network of higher education institutions which promotes transformational learning and change for sustainable development within the higher education sector. Out of its 21 member universities, nine are Austrian.

The United Nations University (UNU) launched its Regional Centres of Expertise on Education for Sustainable Development (RCEs) (UNU-IAS, 2010, p. 21) initiative in 2005. These centres are networks of existing institutions, which work together to promote learning for a sustainable future. They bring together formal education (schools, universities, etc.), non-formal education (museums, nature gardens) and local municipalities, companies and/or non-governmental organizations (NGOs). *'Higher education institutions were especially encouraged to take the lead in developing an RCE because they were expected to provide guidance and leadership in all education and take the initiative to align education from pre-school through university*' (UNU-IAS, 2010, p. 23). In Austria, the University of Graz and the Vienna University of Economics and Business have an RCE.

In summary, it can be said that the university sector has initially not been in the focus of ESD as much as school education, which might be explained by the fact that ESD was originated and driven by a development policy angle. Furthermore, ESD deals by definition with shaping the future by analysing 'the present and studies of the future, and then using these conclusions to take decisions and understand them before implementing them individually, jointly and politically'. (Transfer-21 Programme's 'Quality and Competences' working group, 2007, p. 12). This entails the ability to cope with uncertainty as the notion of an absolute truth is given up. According to Arjen Wals (2010, p. 116), who has conducted the mid-term review of the DESD, 'this [dealing with uncertainty] is a major challenge for higher education as traditionally many scientists consider minimizing uncertainty and maximizing predictability one of their key quests'.

However, universities have come more and more into the focus of ESD since the link between science and society through transdisciplinary research has proven indispensable for a sustainable future.[11] This process has also been reflected in the Austrian context as can be seen in the following.

National dimension

The universities' initiatives on ESD are embedded in a broader context, since the Austrian government has committed to foster sustainable development in different strategic levels. Austria participated in the Earth Summit in 1992 and ratified Agenda 21 at the same event (BMLFUW, 2014a).

In 2002 the Austrian government issued a strategy for sustainable development (BMLFUW, 2002). Even though there exists a broad consensus that education is a key to SD, little of this is mentioned in the strategy, and even less about universities. Out of 20 strategic goals only one deals with education, namely key objective four – 'Solutions Through Education and Research' – which focuses on access, lifelong learning and quality control (sustainability audits) (BMLFUW, 2002, p. 35). Currently, a new version of the national strategy is being drafted and under consultation; the ten main chapters, however, do not specifically mention education (BMLFUW, 2014b). In addition, the Austrian states and the government launched a supplementary strategy (BMLFUW, 2014c) in 2009, which has one out of nine thematic fields focusing on 'Education, Communication and Research for Sustainable

Development'. This strategy is being monitored and has regular work programmes and progress reports.

Concerning the university sector, in 2004 a strategy for research on sustainable development was developed by three ministries, the so-called FORNE (Forschung für nachhaltige Entwicklung).[12] Different research programmes were financed by this initiative which focuses on fields like climate change, energy efficiency and innovative production methods in the first decade of the 21st century. In 2008, a strategy for Education for Sustainable Development[13] was presented. The chapter on higher education mentions FORNE, the Austrian Sustainability Award and a further education programme on ESD for teachers by the University of Klagenfurt.

Apart from political commitments, the Ministry for Science and Research supports concrete projects on ESD in higher education. There are two student initiatives, first the 'Sustainability Challenge', which is an interdisciplinary course on sustainable development at the University of Vienna, the Technical University of Vienna and the Vienna University of Economics and Business with the NGO INEX. 'Go EcoSocial', the other student initiative, is a web-based platform which brings interested students, enterprises and institutions together for master's theses on topics on sustainable development.

The Austrian Sustainability Award[14] was introduced by the Ministry for Science and Research in cooperation with the Federal Ministry for Agriculture, Forestry, Environment and Water Management in 2008. A biennial prize is awarded to higher education institutions in eight categories. The categories are as follows: curriculum and education, research, structural implementation, administration and management, communication and decision-making, European integration, regional integration and student initiatives. An expert jury evaluates the entries in each category and selects the award-winning projects. According to Clemens Mader (2012, p. 88), the Austrian Sustainability Award is the first 'positive attempt for national sustainability appraisal in HEIs' and 'motivates them [the HEIs] to initiate projects and makes them reflect on prior actions for further development'.

Another positive attempt by the Ministry for Science and Research consisted in facilitating internships for 100 adolescents at universities and other research institutions in the field of sustainability research. This initiative was launched in the wake of the Rio+20 conference and has been extended for this year.

As can be seen from above, the Austrian government has shown interest in SD more than ESD by adopting a range of different strategies which hardly mention the educational aspects. The Ministry for Science and Research, however, has been supportive by fostering concrete projects in collaboration with universities.

Institutional dimension (performance agreements 2013–2015)

Die Medizinische Universität Graz steht unter dem Leitgedanken: 'nachhaltig leben.lernen.forschen' (Medical University Graz/Ministry of Science and Research 2012, p. 3); "The Medical University of Graz's guiding idea is 'live.learn.do research sustainably.'", introduction to social goals in the performance agreement of the Medical University of Graz.

For this section, the performance agreements 2013–2015 of the 21 public universities in Austria have been screened to analyse the currently most important aspects in the field of ESD which will be highlighted by the universities within the next three years.

In the latest performance agreements, eight universities have mentioned 'sustainability' or 'sustainable development' in their profile/ mission statement. Out of these, three universities go even further by making sustainability one of the key areas that they have defined for their work. Only four universities have not mentioned sustainability in their performance agreements at all; the rest of them have mentioned 'sustainability' as part of their study programmes, under cooperation projects or merely as a quality indicator in different fields.

Only five universities associate the term with environmental topics in a narrow sense (such as greening the campus or energy-saving initiatives). Many times, sustainability has got to do with cooperation: eight universities have mentioned it in the context of inter-university projects. By means of example, the University of Graz, the Medical University of Graz, the University of Art and Design Linz and the Humboldt University of Berlin have chosen creativity as a major driver for sustainability and will launch a common platform for European creativity research.

The University of Graz, the Medical University of Graz, and Graz University of Technology have founded BioTechMed-Graz, a project in the medical field, which is geared towards sustainability and permanence. It is creating a common cooperative platform for molecular biomedicine, neurosciences and so on. Sustainicum – another cooperation project – has been initiated by the University of Natural Resources and Life Sciences, Vienna, the University of Graz and the Graz University of Technology. The project's goal is to integrate sustainability into university teaching by providing online teaching material.

There are five universities which have centres for sustainability research, two of which (University of Natural Resources and Life Sciences and University of Klagenfurt) have designed complementary study programmes on sustainability. In Graz all four universities have created the platform Sustainability4You as a tool to cooperate in sustainability matters. They organize lectures, work on topics such as mobility together and also use the label 'Sustainable Universities Cluster Graz' for their public relations.

Sustainability projects would typically be mentioned in the chapter on social goals of the universities' performance agreements, hence in a next step this chapter of the performance agreements was analysed. Interestingly, every single university has chosen gender mainstreaming as their main focus in this field. The activities chosen range from daycare centres to proactive assignment of women for professor positions to more technical approaches such as gender budgeting and voluntary participation in 'work-family' audits. The second most often mentioned goal (nine universities) is projects for people with disabilities and comprises barrier-free access to buildings, study materials and proactive hiring policies. Other (less often mentioned) goals are endeavours against discrimination (three universities), diversity management (four), increase in apprenticeships and opening the university to children (four) and schools (three). Interestingly, projects related to widening access to students with a working class background, retired people and migrants have hardly been mentioned (one each).

In 2012, the so-called 'Allianz Nachhaltige Universitäten in Österreich' (Alliance for Sustainable Universities in Austria) was founded.[15] The alliance can be seen as the most comprehensive initiative so far given that nine[16] out of 21 universities are members. The member universities have included identical texts in their respective performance agreements. The alliance's goal is to foster sustainability issues in the university sector and activities will be launched in the fields of knowledge transfer, public relations, sustainability research and so on. The network's actions will ideally be the base of a strategy for sustainability for the Austrian universities. By integrating the network in the performance agreements, it becomes more formalized and the universities have also already mentioned that they are planning on applying for structural funds for their network. These structural funds make up a small percentage of the overall university funding by the government and aim at fostering cooperation between the Austrian universities.

To sum up, the analysis of the performance agreements shows that universities have taken a bottom-up approach in the field of ESD which can be shown by different cooperation projects. These projects reflect different interpretations of sustainability (creativity, teaching methodology, medical research) and thereby give transformative meaning to the policies which have been adopted on a national/international level. On a governance level, sustainability has been increasingly included in the universities' mission statements: whereas in the first performance agreements period (2007–2009) three universities mentioned 'sustainability' or 'sustainable development' in their profile/mission statement, in the second period (2010–2012) it was six universities and in the latest period eight universities. This can be interpreted as an endeavour to governmentalize the bottom-up approaches.

Steering strategies for the future

The last section will provide a meta-analysis of the two change processes and distinguish between top-down and bottom-up strategies. By looking at the three levels of analysis – historical/international, national and institutional – it becomes apparent that the Bologna Process and ESD have followed different implementation strategies. Nevertheless, both processes share the same goal: reforming the Austrian university sector.

Top-down versus bottom-up

The Bologna Process has been implemented through a number of initiatives starting with declarations at the European level (soft law) and followed by reforms of the national university acts to enable universities to change, for example, to the three-cycle structure. In Austria the ministerial communiqués at the European level have been translated into national (hard) law in certain fields. With the ministerial steering tool of performance agreements universities have been further encouraged to reform their systems to align with the Bologna goals. This implementation strategy describes a clear top-down approach and has been adopted by many European governments. This has led to broad critique from the university sector, including student protests and criticism by academics and administrators. Nevertheless, the Bologna Process has turned out to be the decisive change factor for the university system (Byrne et al., 2013, p. 25). Reichert (2010, p. 17) summarizes the reactions to the paradigm shift as follows:

> For many, Bologna has become the symbol of changed higher education attitudes, of a different academic culture and landscape. Remarkably, in several countries the new governance structures were

seen as an erosion of academic freedom, egalitarian values and democratic culture, resulting in controversy from students and academics (for example France and Spain).

In the case of ESD, the initial focus also arose in agreements at international level. However, national implementation remained at the level of soft law and consisted of a variety of ministerial strategies which, typically, have not been supported by concrete action apart from the FORNE strategy which came with a budget for sustainability research. At institutional level, the analysis of the performance agreements has identified different inter-university cooperation projects which have been attributed to sustainability. The variety of the projects shows that the bottom-up approach has left room for interpretation of ESD and the universities themselves have thereby given meaning to ESD in Austria. A bottom-up movement like this requires intrinsic motivation and often includes many hours of voluntary work. Looking at the winners of the Austrian Sustainability Award, numerous projects have come from student initiatives and university personnel. These region- and sector-specific endeavours provide tailor-made solutions but, at the same time, lack the structure and therefore the power of the Bologna Process. It is interesting to note that, at the same time, sustainability has been more and more included in universities' mission statements and thereby 'governmentalized'. So far this process has culminated in the formation of the Alliance for Sustainable Universities in Austria.

Internal and external quality management

Within these top-down and bottom-up approaches internal and external quality management tools are seen as being the main drivers for change within the institutions. Before looking at the different quality management tools the different perspectives on quality in higher education have to be considered (Bernhard, 2012, p. 74):

1. the quality of different spheres in higher education: quality of research, quality of teaching, quality of management and administration
2. the quality of a certain sector of higher education: university sector, non-university sector, post-secondary higher education, tertiary education, higher education
3. the quality at national or international level.

In this chapter the focus is on change processes in the university sector. All three areas of quality in higher education (research, teaching and learning, management) will be covered at the national (in terms of 21 universities) and international level (primarily in terms of organizations operating at the European level which are involved in the relevant change processes). Different internal and external quality management instruments can be seen along the historical dimension, the prevailing situation and the future developments of the observed processes. While internal mechanisms are the core instruments to achieve the goals of the respective transformation processes, external mechanisms give clarity and credibility about the desired results sought by the ministry, other important stakeholders and society.

Both change processes – the Bologna Process and ESD – have seen the interplay of external quality management measures (for example through the use of performance agreements) and internal quality management processes (for example through work load measurement within the Bologna Process or recommendations for ESD being addressed and monitored by the university management). Figure 5.1 shows the different tools that have been identified in our analysis of the two change processes.

In the case of the Bologna Process, a variety of quality management tools are used. Though a common EHEA is an ambitious goal, Austria

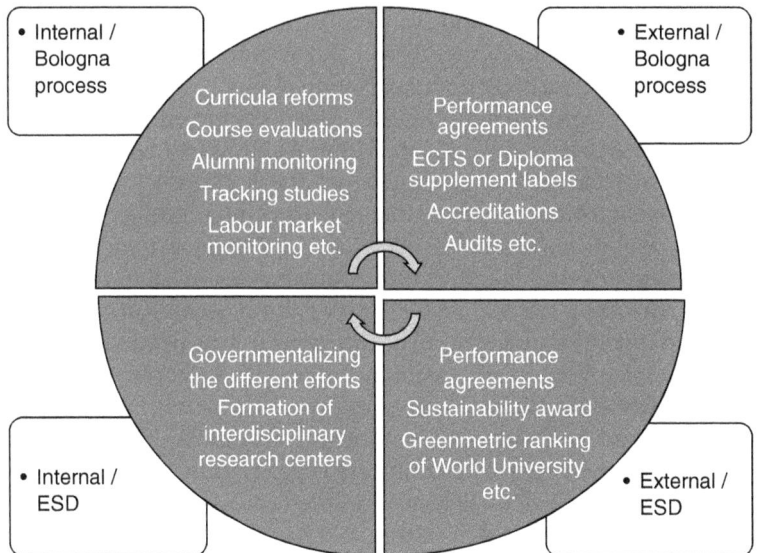

Figure 5.1 Internal and external quality management tools as drivers for change

has performed good implementation efforts. One strategy is to set themes for change. For example, at 'Bologna Day – Quality Management & Quality Assurance in Teaching', organized by the Austrian Agency for International Cooperation in Education and Research (OEAD), which took place in March 2013[17] in Dornbirn, the Bologna expert, Gudrun Salmhofer, identified the main targets for the next years to be 'student-centred learning' and 'learning outcomes'. The guiding questions are to be: how can universities enact these popular concepts with content and substance, how can universities be supported to implement them in a better way, and how can universities be convinced to start innovations in these areas. Furthermore, Salmhofer states that the Bologna Points which had been previously established at universities were renamed because of market-oriented interests (for example Centre for Teaching and Learning at the University of Vienna) as Bologna is blamed for bringing burdens, complications and more chaos instead of clarity and improvement for universities.

Concerning the issue of quality assurance within the Bologna Process, Achim Hopbach, managing director of the Agency for Quality Assurance and Accreditation Austria (AQ Austria), noted at the 'Bologna Day' that quality assurance has been part of sector stocktaking since 2009 and that this will continue until 2015 when, in Yerewan, the European Standards and Guidelines (ESG) (ENQA, 2009) will be revised. Currently, programme accreditations and evaluations are widely spread throughout Europe and institutional accreditation and evaluation along with audits are becoming increasingly popular. National quality assurance systems are implemented in line with the ESG. The development of a 'uniform' European quality assurance system is improbable as taking into account the national context is important. From the Bucharest conference in 2012 a more holistic approach to implementing interdependent Bologna tools has been promoted (EHEA, 2012, p. 3): 'The development, understanding and practical use of learning outcomes is crucial to the success of ECTS, the Diploma Supplement, recognition, qualifications frameworks and quality assurance – all of which are interdependent.'

The central role of quality assurance processes in implementing the Bologna goals is now clear. However, questions about how best to design an efficient and productive external quality assurance system and how to improve student-centred learning are yet to be fully answered.

Looking at the use of quality management tools within ESD, we note that the bottom-up approach in this field has also been reflected in the development of relevant quality assurance tools. For example, external quality assurance tools like the Austrian Sustainability Award, national sustainability reports and the Greenmetric World University Ranking[18] have been undertaken voluntarily. These endeavours might have historically also served the purpose of a better visibility of ESD initiatives. By applying these and other strategies, some universities have reached a position where sustainability has become a leading principle and has been included in their mission statement. The Alliance for Sustainable Universities in Austria now plans to define what quality means in the field of ESD in Austria and what emerges can be supported into implementation using the range of internal and external quality assurance and change management tools now available.

Concluding remarks

Our case study has identified that, in response to the question *how change processes take place at universities in Austria*, it is driven by a mixture of top-down and bottom-up strategies in combination with a range of different internal and external quality management processes.

Compared to the Bologna Process, ESD in the university sector has been evolving at a slow pace. This can be explained by its implementation being fostered using soft law (recommendations, guidelines and governmental strategies) whereas the implementation of the Bologna Process has been fostered using hard law (University Act 2002, Amendment Act 2009). The implementation of the Bologna Process (for example three-cycle structure, qualification profiles) has been supported by the establishment of Bologna Points, projects of the uniko and also inter-university exchange of best practices.

In the case of Bologna universities have started to select their own emphases on certain topics like the third cycle, tracking studies and quality assurance in order to give stronger meaning to the Bologna Process. Given the more noncommittal nature of ESD, meaning has been defined mainly by dedicated university members. This has led to a more diverse approach within the university sector. The Alliance for Sustainable Universities in Austria can be a platform to discuss a more integrated and focused approach to ESD and assist this by suggesting a common quality approach. However, ultimately quality in the area of ESD can only be defined by the institutions themselves: a process which also reflects the European trend towards quality audits.

It is difficult to find the right strategy to initiate change processes but a mixture of top-down and bottom-up ones has supported the implementation of both the Bologna Process and ESD. Since every process is unique by nature and involves different university contexts and stakeholders, there is no such thing as a one-size-fits-all solution for success. However, it is important to agree on a common vision by including the people who will actually put it into practice to help shape it. We want to end with Staudacher's (2012, p. 77) analysis of the status of Austrian universities within the Bologna Process that the good news is: 'We are on the right track!'

Notes

1. An English version of the University Act is available online at the website of the Federal Ministry of Science, Research and Economy: http://wissenschaft.bmwfw.gv.at/fileadmin/user_upload/legislation/E_UG.pdf
2. A more detailed description of change processes at Austrian universities linked to the Bologna Process can be found in 'Transformation 1: Bologna Process'.
3. All performance agreements are publicly available on the uni:data warehouse of the Federal Ministry of Science, Research and Economy: http://www.bmwfw.gv.at/unidata (German only).
4. German word for 'sustainable' which also shows all hits for sustainability as it is part of the word 'nachhaltigkeit'.
5. A more detailed description of performance agreements can be found in 'Method of Analysis and Structure'.
6. The monitoring reports can be downloaded at the Website of the Ministry of Science, Research and Economy: http://wissenschaft.bmwfw.gv.at/bmwfw/studium/studieren-im-europaeischen-hochschulraum/bologna-prozess/der-europaeische-hochschulraum-im-oesterreichischen-kontext/ueberpruefung-der-bologna-umsetzung/bologna-monitoring-report/
7. Universities Austria is the Austrian Rectors' Conference which is a non-profit association under private law. Its purpose is to assist the Austrian universities in the fulfilment of their tasks and responsibilities and thus to foster scholarship and research. Universities Austria handles the internal coordination of the 21 public Austrian universities, it represents them in national and international organisations and is the public voice of the universities. For more information see www.uniko.ac.at.
8. Projects related to the Bologna Process by Universities Austria: http://www.uniko.ac.at/arbeitsbereiche/lehre/schwerpunkte/bologna_prozess/
9. Chair from the national BFUG and head of the Bologna contact point at the Ministry of Science, Research and Economy.
10. For more information see http://www.copernicus-alliance.org
11. In the literature this change is described as a move from Mode 1 towards Mode 2 of knowledge production (Gibbons et al., 1994; Nowotny et al., 2003).
12. German for 'research for sustainable development'.
13. An English short version of the Austrian Strategy for Education for Sustainable Development is available on http://www.bmukk.gv.at/medienpool/18300/bine_strategie_E.pdf
14. For more information see: http://www.bmwf.gv.at/startseite/wI_und_fo_gemeinsam/nachhaltigkeit (German only)

15. It is the only project on sustainability mentioned under 'social goals' and was included by eight of the nine member universities as a social project.
16. University of Graz, Graz University of Technology, University of Innsbruck, University of Klagenfurt, Vienna University of Economics and Business, University of Natural Resources and Life Sciences, Medical University of Graz, University of Music and Performing Arts Graz and University of Salzburg
17. For more information about Bologna Day see: http://www.bildung.erasmusplus.at/hochschulbildung/europaeischer_hochschulraum/veranstaltungen_trainings/bologna_tag/ (German only).
18. The Greenmetric World University Ranking (http://greenmetric.ui.ac.id/) is an initiative by the Universitas Indonesia, which focuses on the ecological footprint of universities. Several Austrian universities have taken part.

References

APA (Austrian Press Agency) (2012) *Schmidinger: Uni-Autonomie täglich neu Erkämpfen* (press release 2 October 2012), http://www.uniko.ac.at/modules/download.php?key=2388_DE_O&f=1&jt=7906&cs=5D87, date accessed 15 April 2014.

Bacher G. (2013) 'Was gibt es Neues im EHR?', http://www.bildung.erasmusplus.at/fileadmin/oead_zentrale/projekte_kooperationen/bologna-service/Bologna_Tag_2013/plenum_bfug.pdf, date accessed 15 April 2014.

Bernhard A. (2012) *Quality Assurance in an International Higher Education Area. A Case Study Approach and Comparative Analysis* (Wiesbaden: VS Verlag für Sozialwissenschaften).

BMLFUW (Ministry of Agriculture, Forestry, Environment and Water Management) (2002) 'The Austrian Strategy for Sustainable Development', https://www.nachhaltigkeit.at/assets/customer/Downloads/Strategie/strategie020709_En.pdf, date accessed 15 April 2014.

BMLFUW (Ministry of Agriculture, Forestry, Environment and Water Management) (2014a) 'Lokale Agenda 21 in Österreich', http://www.bmlfuw.gv.at/umwelt/nachhaltigkeit/lokale_agenda_21/lokaleagenda21oest.html, date accessed 15 April 2014.

BMLFUW (Ministry of Agriculture, Forestry, Environment and Water Management) (2014b) 'Erneuerung der NSTRAT in Arbeit', https://www.nachhaltigkeit.at/strategien/strategie-des-bundes/erneuerung-der-nstrat-steckengeblieben, date accessed 15 April 2014.

BMLFUW (Ministry of Agriculture, Forestry, Environment and Water Management) (2014c) 'ÖSTRAT – Nachhaltigkeitsstrategie des Bundes und der Länder', http://www.bmlfuw.gv.at/umwelt/nachhaltigkeit/strategien_programme/oestrat.html, date accessed 15 April 2014.

Bundeskanzleramt-RIS (2002) 'Bundesgesetz über die Organisation der Universitäten und ihre Studien (Universitätsgesetz 2002 – UG', http://www.ris.bka.gv.at/Dokumente/BgblPdf/2002_120_1/2002_120_1.pdfhttp://www.ris.bka.gv.at/Dokumente/BgblPdf/2002_120_1/2002_120_1.pdf, date accessed 15 April 2014.

Byrne J., Jørgensen T., Loukkola T. (2013) 'Quality Assurance in Doctoral Education. Results of the ARDE Project' (EUA Publications 2013), http://www.

eua.be/Libraries/Publications_homepage_list/EUA_ARDE_Publication.sflb.ashx, date accessed 15 April 2014.
Eaton J. S. (2001) 'Regional Accreditation Reform: Who Is Served?' *Change Magazine*, 33(2) 38–45.
EHEA (2005) 'The European Higher Education Area – Achieving the Goals', communiqué of the conference of European ministers responsible for higher education, Bergen, 19–20 May 2005.
EHEA (2012) 'Making the Most of our Potential: Consolidating the European Higher Education Area', Bucharest communiqué, Bucharest, 26 and 27 April 2012.
ENQA (European Association for Quality Assurance in Higher Education) (2009) *Standards and Guidelines for Quality Assurance in the European Higher Education Area*, 3rd edn (Helsinki: ENQA).
European Commission (2014) 'Europe 2020: Priorities', http://ec.europa.eu/europe2020/europe-2020-in-a-nutshell/priorities/index_En.htm, date accessed 15 April 2014.
Gibbons M., Limoges C., Nowotny H., Schwartzman S., Scott P., Trow M. (1994) *The New Production of Knowledge: The Dynamics of Science and Research in Contemporary Societies* (London: Sage Publications).
Loprieno A., Menzel E., Schenker-Wicki A. (2011) 'Zur Entwicklung und Dynamisierung der österreichischen Hochschullandschaft – eine Außensicht: Rahmenkonzept für einen Hochschulplan', http://hochschulplan.at/wp-content/uploads/2012/06/Bericht_ExpertInnen_2011.pdf, date accessed 15 April 2014.
Mader C. (2012) 'How to Assess Transformative Performance towards Sustainable Development'. *Journal of Education for Sustainable Development*, 6(1) 79–89.
Medical University Graz/Ministry of Science and Research (2012) 'Leistungsvereinbarung 2013–2015', http://www.meduni-graz.at/images/content/file/organisation/grundsatzdokumente/LV_2013–15.pdf, date accessed 15 April 2014.
Ministry of Science and Research (2012) 'Bologna Monitoring: Bericht über den Stand der Umsetzung der Bologna Erklärung in Österreich. Berichtszeitraum 2010–2012', http://wissenschaft.bmwfw.gv.at/bmwfw/studium/studieren-im-europaeischen-hochschulraum/bologna-prozess/der-europaeische-hochschulraum-im-oesterreichischen-kontext/ueberpruefung-der-bologna-umsetzung/bologna-monitoring-report/, date accessed 15 April 2014.
Nowotny H., Scott P., Gibbons M. (2003) 'Introduction. "Mode 2" Revisited: The New Production of Knowledge'. *Minerva*, 41, 179–194.
OECD (2011) *Better Policies for Development: Recommendations for Policy Coherence* (Paris: OECD Publishing).
Oxford Dictionaries (2014) 'About Oxforddictionaries.com', http://www.oxforddictionaries.com/words/about, date accessed 15 April 2014.
Pichl E. (2012) 'Universitäre Profilbildung im Kontext des österreichischen Hochschulraums und des Universitätsgesetzes 2002'. *Zeitschrift für Hochschulrecht, Hochschulmanagement und Hochschulpolitik*, 6, 195–206.
PROVISO (2013) '7. EU-Rahmenprogramm für Forschung, technologische Entwicklung und Demonstration (2007–2013), PROVISO-Überblicksbericht (Datenstand 11/2013)', http://wissenschaft.bmwfw.gv.at/fileadmin/user_

upload/proviso/PROVISO_EB7rp3550eha140514_Gesamt_Druck.pdf, date accessed 15 April 2014.
Reichert S. (2010) 'The Intended and Unintended Effects of the Bologna Reforms'. *Higher Education Management and Policy*, 22 (1) 1–20.
Skordoulis R. T. (2012) 'Change Management in Higher Education: Top-Down or Bottom-Up?' *International Journal of Applied Management Education and Development*, 1(3) 1–19.
Statistik Austria (Austrian Statistical Agency) (2013) *Bildung in Zahlen 2011/12 – Schlüsselindikatoren und Analysen* (Wien: Statistik Austria).
Staudacher E. (2012) 'Bologna Reloaded – Wo müssen wir nachladen?' *Zeitschrift für Hochschulrecht, Hochschulmanagement und Hochschulpolitik*, 11 (2) 76–85.
Sursock A., Smidt H. (2010) *Trends VI: A Decade of Change in European Higher Education* (Brussels: European University Association).
Transfer-21 Programme's 'Quality and Competences' working group (2007) *Guide: Education for Sustainable Development at Secondary Level Justifications, Competences, Learning Opportunities* (Berlin: Transfer-21 Programme Koordinierungsstelle Freie Universität Berlin).
Universities Austria (ed.) (2008) *Recommendations by Universities Austria on New-Style Doctoral Studies* (Vienna: Universities Austria).
UNU-IAS (2010) *Five Years of Regional Centres of Expertise on ESD* (Yokohama: UNU-IAS).
Wals A. E. J. (2010) 'DESD We Can? Some Lessons Learnt from Two Mid-DESD Reviews'. *Global Environmental Research*, 14, 109–118.
Weiler H. N. (2006) 'Profil – Qualität – Autonomie. Die unternehmerische Universität im Wettbewerb'. *Zeitschrift für Hochschulrecht, Hochschulmanagement und Hochschulpolitik*, 2, 39–46.
Westphal E. (2010) 'Kommentar. Halbzeit auf der Bologna-Reise: Impressionen aus der ersten Dekade'. In Universities Austria (ed.) *Jahresbericht 2010* (Vienna: Universities Austria).

6
A Quality Assurance System Based on the Sustainable Development Paradigm: The Lithuanian Perspective

Laima Galkute

Introduction

In contemporary societies, higher education is is an important sector because of its increasing role in negotiating the complex, interlaced issues of development and involvement in designing and implementation corresponding strategies at the national and local level. Changing societal expectations in Lithuania, along with international influences arising from the European Higher Education Area (EHEA) and European Union (EU), have a significant impact on national higher education policy. In this context a quality assurance (QA) system is considered a sensitive and influential policy instrument by reflecting both the relationship between the state and higher education institutions (HEIs) and between their autonomy and responsibility for societal transformation.

In this chapter the emerging model of QA in Lithuanian higher education is discussed by exploring its interplay with the sustainable development paradigm.

Changing society through changing education

The Bologna Declaration (EHEA, 1999) initiated an international process in order to create the common Higher Education Area by coherent reforming of national higher education systems to fulfil their diverse missions in a knowledge society. The original document was signed by 30 European countries; there are now 47 participating countries, not only within Europe, but worldwide. The Bologna Process was supported by a series of meetings of the European ministers responsible for higher education (EHEA 2001, 2003, 2005, 2007, 2009, 2012a). Policy decisions have been taken in the areas of curriculum reform and degree structure, recognition of qualifications, lifelong learning, international openness and mobility, the social dimension of higher education, as well as in QA.

When considering QA from the point of view of its transformative capacity, it is interesting first to understand the role of higher education as it is reflected in the outcomes of ministerial meetings (Table 6.1). Starting with structural changes for harmonizing higher education systems across EHEA, the current policy focus is on innovation-driven societal development and building the intellectual independence and self-confidence of students. The scope of the HEI activities under consideration is also evolving from study programmes towards the whole institution approach.

In the Graz Declaration 2003, the European Universities Association (EUA, 2003) defined the role of universities as being to 'create, safeguard and transmit knowledge vital for social and economic welfare, locally, regionally and globally'. The transformation of HEIs was emphasized by advocating the integration of reforms 'into the core institutional functions and development processes, to make them self-sustaining'. The Graz Declaration 2003 was presented to the EHEA ministers in their meeting in Berlin.

In the European Union, HEIs are recognized as the key players for ensuring the successful transition to a knowledge-based economy and society, which they do by strengthening partnerships with business (Europa, 2006). The modernization agenda for national higher education systems up to 2020 (Europa, 2011) at the policy level calls for systematic involvement of HEIs in preparing and implementing local and regional development plans, particularly through the creation of regional hubs of excellence and specialization. In the long run, the modernization agenda supports the implementation of 'Europe 2020, a Strategy for Smart, Sustainable and Inclusive Growth' (Europa, 2010a), targeted at a high quality of healthy life, underpinned by Europe's unique social models.

With respect to sustainable development specifically, it was emphasized in the 'Council Conclusions' (Europa, 2010b) that educational institutions at all levels should act as role models, by integrating the principles of sustainable development into policy and practice, including energy-saving, building and working with natural resources and sustainable purchasing and consumer policy.

Table 6.1 Changing rhetoric on the role of higher education as reflected in the EHEA ministerial meetings

Changing rhetoric in the EHEA as reflected in the ministerial meetings	
Prague, 2001	...study programmes combining academic quality with relevance to lasting employability.
Berlin, 2003	...increase the role and relevance of research to technological, social and cultural evolution and to the needs of society.
Bergen, 2005	...the importance of research in underpinning higher education for the economic and cultural development of our societies and for social cohesion.
London, 2007	...important influence higher education institutions (HEIs) exert on developing our societies, based on their traditions as centres of learning, research, creativity and knowledge transfer as well as their key role in defining and transmitting the values on which our societies are built.
Leuven/ Louvain-la-Neuve, 2009	...public policies will fully recognise the value of various missions of higher education, ranging from teaching and research to community service and engagement in social cohesion and cultural development.
Bucharest, 2012	Our societies need higher education institutions to contribute innovatively to sustainable development and therefore, higher education must ensure a stronger link between research, teaching and learning at all levels.

Lithuanian reform of higher education and its role for sustainable development

The Bologna Process stimulates transformation of higher education policies in all countries within the EHEA. In Lithuania it has triggered changes in degree systems and the relationship between higher education and research, and stimulated academic mobility and the development of national QA systems. All these new developments were formalized in the Law on Higher Education and Research (Seimas, 2009).

According to the Law, 'A coherent system of higher education and research is the foundation of the development of a knowledge society, the strengthening of a knowledge-based economy and the sustainable development of the country.' The perspective up to 2020 is presented in the State Programme for Studies, Research and Experimental (Social, Cultural) Development (Government, 2012b). It aligns well with the Europe 2020 strategy and, at the national level, with the National Development Programme for 2014–2020 (Government, 2012a), which gives focus to promoting sustainable development as a horizontal principle for all sectors.

The State Programme for Studies emphasizes three major goals: (1) widening access to higher education irrespective of the social status of students, while, at the same time, enhancing the proportion of students in the physical and technological sciences; (2) creating new knowledge and stimulating the integration of research, business and culture; (3) ensuring relevant and effective management of higher education including by putting in place robust QA systems, fostering openness and accountability as well as the leadership and innovation capacity of HEIs. It should be noted that the development of internal QA systems is particularly encouraged because of the shift to external evaluation of internal QA systems.

At the institutional level, rethinking of the role and profile of HEIs towards sustainable development is recognized as well. Some Lithuanian examples of the HEI mission statements which are being used as a basis for their long-term strategies are presented in Box 6.1.

Box 6.1 Examples of university mission statements

Kaunas University of Technology

The mission of the university is to provide research-based studies of international level, to create and to transfer knowledge and innovative technologies for the sustainable development and innovative growth of the country, to provide an open creative environment that inspires leaders and talented individuals.

Source: www.ktu.edu.

Klaipeda University

Klaipeda University is a centre of Lithuania as a marine country and a centre of research, arts and studies in the Baltic Sea region, which prepares highly qualified specialists, fosters humanist values and pays parallel priority attention to: research in marine science and marine studies; history, culture and languages, education, health and social welfare, economy, politics, communications and the arts of the Baltic Sea region; sustainable development of Western Lithuania and the city of Klaipeda; development of an integrated science, studies and business centre.

Source: www.ku.lt.

Aleksandras Stulginskis University (University of Agriculture)

The mission of the university is to create and disseminate scientific knowledge and strive sincerely for safe and healthy food and a full-fledged living environment for all people of Lithuania.

One of the steps in meeting this major goal includes the development and dissemination of biological, engineering and social technologies, and advanced knowledge and experience in the sphere of sustainable use/development of land, forest and water resources.

Source: www.asu.lt.

In higher education, the involvement of employers is promoted both by the European and national strategic documents. A consensus among politicians and various stakeholders (teachers, students, employers) is extremely important both in determining strategic priorities for the HEIs of Lithuania and in formulating the quality criteria to be used to evaluate implementation.

In 2012 a national survey including questions on the role of higher education was organized by the Research and higher education monitoring and analysis centre at the Lithuanian Ministry of Education and Science. The survey covered three target groups: (1) university teachers (1004 respondents); (2) HEI administrators (1003 respondents); (3) employers involved in a long-term partnership with specific HEIs (968 respondents).

The survey was carried out by means of a face-to-face standardized interview. Quota sampling in the case of teachers was used to ensure responses were representative of the type of HEI (university and college/university of applied sciences) and discipline area (humanities, arts, social sciences, physical sciences, biomedical sciences and technological sciences). Figure 6.1 shows the distinctive opinions in defining the role of higher education which emerged from the survey.

Prior to discussing Figure 6.1, it is necessary to mention that all of the roles identified are important from the point of view of sustainable development. However, the prevalent focus on preparing professionals for the labour market indicates outgoing understanding which is characteristic for 'industrial society', while the focus on a more comprehensive set of competences shows a shift towards late modernity and a 'creative society'.

The survey identified significant (and unexpected) differences between university teachers and employers in their understanding of the role of higher education. Employers gave much more emphasis to the needs of the labour market compared with teachers (Figure 6.1). 'Internal' differences between teachers in universities and colleges were also found (not shown in Figure 6.1). Supporting the labour market was more important in colleges (this is understandable given their key role preparing profession-oriented graduates) than in universities: 31.8 per cent and 17.1 per cent respectively, in common – 24.4 per cent. Universities indicated more interest in innovation: 8.9 per cent for universities and 5.2 per cent for colleges (in common –7.1%). The opinions of university administrators were quite close to the teachers.

There could be many explanations for these different importance ratings. However, the key implication is that there needs to be far more systematic discussion between key stakeholders and in society at large on both the role and quality of higher education, including its role in fostering a sustainable future.

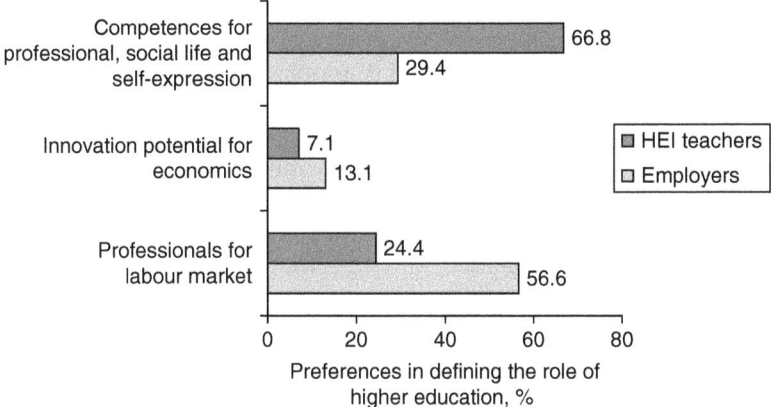

Figure 6.1 Preferences in defining the role of higher education by university teachers and employers

A focus on quality assurance

In the EHEA, 'quality assurance in higher education can be understood as policies, procedures and practices that are designed to achieve, maintain or enhance quality as it is understood in a specific context' (EHEA, 2012b, p. 60).

From the outset of the Bologna Process, it was agreed that each HEI should be responsible for QA thereby reflecting the principle of institutional autonomy (EHEA, 2003). At the same time, it was also emphasized that it was necessary to develop mutually shared criteria and methodologies for QA. The Standards and Guidelines for Quality Assurance in the European Higher Education Area (ESG) were established in 2005. These give focus to the interplay between three components of the QA system: (1) internal QA within the HEI; (2) the external QA of higher education; and (3) operation of external QA agencies (ENQA, 2005).

Managing the interplay between internal and external QA and the associated balance between autonomy and accountability in ways that are consistent with each institution's mission poses significant challenges for each HEI. This is why managing institutional diversity and creativity supported by internal quality processes was identified as one of the key challenges at the beginning of the second decade of the Bologna Process (Sursock and Smidt, 2010).

The European University Association, in particular, promotes creativity and innovative practices within European universities through projects which enhance the development of a quality culture and efficient and adequate governance in the context of institutional diversity (EUA,

2005, 2006, 2009; Loukkola and Zhang, 2010; Sursock, 2011; Vettori, 2012). As it was concluded in one of project reports, 'when quality is understood as transformational and fit-for-purpose..., quality assurance should link creativity and quality rather than see them as being mutually exclusive' (EUA, 2009, p. 16).

In spite of seeking to implement the common QA standards and guidelines, a review of the national QA systems currently under way shows diversity in their orientation, whether the main focus of QA is on institutions or programmes, or both (EHEA, 2012b). Some countries are strongly oriented towards evaluation of the internal QA systems, for example, Austria, Finland, the United Kingdom and the Netherlands. The review also notes that QA systems are becoming more complex as they evolve. In relation to internal QA systems, an essential question is about institutional strategies for continuous quality improvement.

The external QA process in Lithuania was introduced in 1995 with the establishment of the Centre for Quality Assessment in Higher Education (hereinafter, the Centre). The main task was to organize the quality assessment of academic activity in HEIs, both of study programmes and research. Since 1999, external evaluation of study programmes at the bachelor's and master's levels, involving foreign experts, has taken place on a regular basis. The commission for the evaluation of Lithuanian universities' research results was established in 2001 to ensure their compliance with master's degree studies. In 2004, the Centre started to carry out institutional evaluations of colleges (universities of applied sciences) as a procedure for introducing a binary system in higher education. Recent developments of QA in Lithuania associated with the higher education reform are presented in the next section.[1]

Approaches and principles for institutional quality assurance in Lithuania

The higher education reform in Lithuania initiated a shift towards the enhanced autonomy of the HEIs, including encouraging flexibility in strategic planning while, at the same time, allocating to them stronger responsibility for assuring the quality of all activities. The new Law on Higher Education and Research (hereinafter, the Law) (Seimas, 2009) determines autonomy and QA for the HEIs as follows:

- Article 7. Autonomy and Accountability of Higher Education Institutions stipulates the right to choose study fields and forms as well as research and development activities (para 2.1), and the obligation to

inform the founders and the public about the QA measures being used in studies and research as well as about the results of external quality evaluation and accreditation of their study programmes (para 3.2);
- Article 40. Quality Assurance in Higher Education and Research stipulates the responsibility of the HEI for the quality of studies, research activities (including in artistic areas) and other activities (para 1) and their continuous improvement based on self-assessment and external evaluation (para 5);
- Article 41. Internal Quality Assurance in Activities of Higher Education and Research Institutions stipulates that every HEI must have an internal system of QA for studies based on ESG and its own strategy for QA (para 1) as well as providing comprehensive quantitative and qualitative information about activities and QA measures (para 2).

According to the Law, the procedures for QA should be defined in the HEI statute (Article 28, para 2.11). The senate of a university or the academic council of a college (the highest self-governance bodies in these sectors) shall approve an internal QA system and control its implementation (Article 21, para 2.3 and para 3.3). It should be mentioned that, being budgetary institutions, the public HEIs are obliged to annually report financial data related to their studies and research.

Changes in legislation motivate for rethinking the objectives and approaches adopted in internal QA processes as well as the interplay between external and internal QA procedures. For example, external expertise and accreditation of study programmes was limited to ongoing ones, whereas approval of intended study programmes was made the responsibility of the individual HEI from 2011.

'Transformation is change that is profound, radical, and sustainable; change that fundamentally and indelibly alters the very nature of something. Not all change is or should be transformational' (Gass, 2012, p. 21). Teaching and learning, research as well as outreach activities which are mutually interdependent should be driven by strategic management in a holistic way. The whole institution approach seems to be the only way to develop robust, internal QA systems and achieve the synergy sought between the HEIs and societal goals at the local, national and European level.

External institutional evaluation using expert review was launched in 2010 for all HEIs, whether they be public or private. It is based on qualitative self-evaluation accompanied by quantitative evaluation of research (art) output and statistics on human resources and infrastructure. Being evidence-based, quantitative assessment is not sufficient for understanding and evaluating complex academic processes which, in turn, are integrated in a variety of development strategies and contexts.

The main documents regulating institutional QA are: 'Procedure for the External Review of Higher Education Institutions', 'Accreditation Procedure of Higher Education Institutions' (Government, 2010) and 'Methodology for Conducting an Institutional Review in Higher Education' (hereinafter, Methodology) (Centre, 2010a). Autonomy and accountability, contextuality, taking a holistic approach, effective stakeholder involvement, unity of internal and external quality assurance, and continuity are defined as the driving principles of the assessment.

The Methodology's principle of taking a *holistic (systemic) approach* at the institutional level is demonstrated in Figure 6.2. It illustrates the interconnection between strategic management of the HEI and internal and external quality assessment based on agreed quality criteria (Box 6.2).

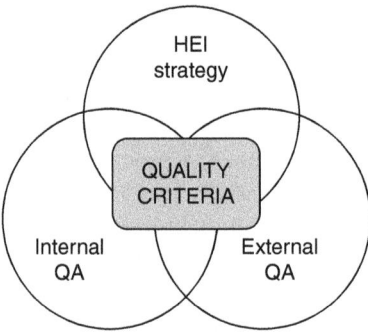

Figure 6.2 Interconnection between strategic management of the HEI and quality assurance

---

Box 6.2 Quality criteria

'A set of criteria that provokes thinking and action regarding quality enhancement rather than quality control. Quality criteria is often created jointly by stakeholders and is open to participatory and ongoing debate. Quality criteria can be considered as a 'translation' of a shared set of stakeholder values prepared in a transparent manner with a practical function'.

Source: UNESCO, 2007, p. 104.

When quality criteria are applied during the review all claims made against them should be supported by qualitative and quantitative indicators.

To be really transformative, internal QA should be integrated into the strategic management cycle of the HEI as a key element. Actually, linking of QA procedures with the institutional strategic cycle and the external evaluation cycle was suggested by the EUA report (Sursock and Smidt, 2010) in order to reduce the financial burden and the time spent on these processes. The new aspect is that, for transformative quality management to work, the internal quality criteria should be coherent with strategic goals of the HEI on the one hand, and with external evaluation on the other, thus reflecting a correspondence to public policy.

An institutional quality review of an HEI in Lithuania covers the following evaluation areas defined as: (1) Strategic management; (2) Studies and life-long learning; (3) Research and/or art activities; (4) Impact on regional and national development. It is noted in the Methodology that it is necessary to analyse and evaluate the interconnection between Strategic management and the rest of the evaluation areas (para 7). It also implies that only a 'basket of indicators' can characterize the dimensions of the HEI development in a comprehensive manner.

Different types of qualitative indicators are integrated into the Methodology – context, input, process and output. It means that analysis is carried out using the 'time' and 'space' perspective (Flick, 2006) that is characteristic of qualitative research:

- The dynamics of qualitative changes (emerging qualities) are reflected in the input-process-output indicators.
- There is comparative assessment of the extent to which internal changes align with the external context (for example, 'compliance of the strategic documents relating to studies and life-long learning with the provisions of the EHEA and the EU documents related to higher education': Centre, 2010a).

The principle of *contextuality* cannot be considered separately from the holistic approach. The Methodology stipulates the obligation of each HEI to set itself quantitative and qualitative indicators to support common quality criteria (para 5.5). Therefore, an external QA is sufficiently flexible to support the diversification of the HEIs' activities in accordance with their strategies and integration into societal processes. There are 30 indicators in total for external evaluation, 13 of which focus on the area of Strategic management. It should be noted that HEIs may use additional indicators if they wish to better represent the implementation of their strategies.

A principle of *stakeholder involvement* was stimulated by the development of the EHEA as a precondition of accessibility and social dimension of higher education as well as employability of graduates. On the other hand, representatives of the main stakeholders (students, academics, employers) usually are involved in the procedures of self-evaluation and external review. The Methodology gives more emphasis to partnerships with academia outside the HEI and with a variety of stakeholders from public and business institutions. It is important to develop the transformative capacity of the HEI by exploring 'feedback impact' from the partners on research and learning as well as by analysing experiences of collaboration in outreach activities.

The principle of *continuity* and learning is at the core of the Methodology: this requires the HEI to demonstrate the capacity for critical evaluation of its activities and envision relevant and feasible strategies for quality improvement (para 15). All the statements in the self-assessment report should be based on quantitative and qualitative evidence. Special attention is paid to the analysis of effectiveness and efficiency of the institution's internal QA system. After the external review, the HEI is to determine the measures for the elimination of the drawbacks identified during the self-evaluation and external review for the improvement of its overall activities (para 54).

However, the most interesting question about the Methodology is not 'what it is?' but 'how does it work?' There are 23 universities in Lithuania; 14 of them are public. Both self-evaluation and external expert reviews were carried out for 13 universities over 2011–2013; the rest will be evaluated during the years leading up to 2015. The institutional review reports are publicly available on the website of the Centre for Quality Assessment in Higher Education both in Lithuanian and in English.

Analysis of the review reports shows an important difference between public and private universities, in that the latter group did not consider the 'third mission' (a key mechanism for identifying and addressing key local social, cultural, economic and environmental issues) as an important component of their activities. Public universities are characterized basically by knowledge transfer related to their profile (educational sciences, sports, agriculture, music and theatre, arts) or by their focus on the various issues in development of particular regions (Klaipeda University and Siauliai University are the distinct examples). These activities represent an object of internal quality monitoring. However, projects in non-formal education, protection of nature and cultural heritage, sustainable use of resources, support to social groups with special needs, and voluntary activities by students usually are interpreted as 'services to community', which are not supported by comprehensive performance indicators.

The institutional self-evaluation exercise is important in identifying and assessing the quality of the actions undertaken by the university in a systemic way. It has also stimulated a shift from quality management (both external and internal) focused mainly on studies and research towards the whole institution approach.

Sustainable transformation as a 'moving target'

Quality assurance, as interpreted and discussed in the previous sections, could be considered as a focal point for finding synergies between the state and the HEI or, in broader scope, between the development dynamics of society and academia. What characteristics are the most important for the efficient functioning of the internal QA system?

Analysis of internal QA systems shows that they often include elements of ISO and/or European Foundation for Quality Management (EFQM) model and are concentrated mainly on studies and research; some universities, however, have introduced the use of the Global Compact for promoting social responsibility. However, these do not include the strategic management for the complex assessment and development of the entire institution.

QA systems targeted towards quality control which were used at the start of the quality movement are less applicable in the current period of late modernity. Contemporary society is characterized by complexity, interconnectedness, rapid change as well as by uncertainty and a wide range of risks encompassing different spheres of life. Such a situation requires a shift from quality control towards dynamic quality research and futures thinking. The Methodology (Centre, 2010a) seems to be a signal for the HEIs to essentially reconsider the notion of internal QA against this context.

The present situation in internal QA systems could be considered to be a consequence of focusing on the ESG which mainly emphasizes characteristics of studies: 'Institutions should have a policy and associated procedures for the assurance of the quality and standards of their programmes and awards' (ENQA, 2005). However, to be really transformative, the QA system should be a vehicle and instrument for strategic management, encompassing all spheres of the HEI's activities (learning and teaching, research, engagement and operations) in a more holistic and integrated way. It should also be responsive to the key challenges of its region, to policy changes that affect it, to the requirements of the external QA process and the expectations and needs of its key stakeholders. Finally, QA should consider dynamics of the societal and academic processes and be flexible and future-oriented in their strategic objectives and priorities.

Lithuanian HEIs are becoming increasingly involved in sustainable development strategies and initiatives. Some of them see sustainable development as an element of their study programmes or greening their campus, others consider the area as a core of their missions and long-term strategies. Universities have an ideal opportunity to 'practise what they preach' in this regard by becoming living laboratories for learning about, researching and demonstrating sustainability in action. As Gass (2012, p. 23) observes, 'The process of transformational change must always model what it seeks to create.' Therefore, the next step is to find ways to facilitate the transition towards sustainable development.

Sustainable development is not a scientific concept but rather it is a general political commitment or a meta-policy with a particularly broad and cross-cutting scope (O'Toole, 2004). It commits society to a long-term engagement, during which changing cognitive circumstances, changing empirical circumstances, and persisting uncertainties should be considered as a permanent condition. However, there are some fundamental principles which need to be considered in orienting QA towards sustainable development (Waas et al., 2011): (1) normativity; (2) equity; (3) integration; and (4) dynamism. The principles of integration and dynamism are actually reflected in the concept of QA under consideration. In addition, a dynamics of internal (bottom-up) and external (top-down) QA systems should be particularly emphasized as it reflects interplay of autonomy and accountability of the HEIs. However, principles of normativity and equity in respect to higher education should be further discussed.

The *equity* principle in higher education has links to the concepts of social and cultural sustainability and development. It is related mainly to the notion of opening up access to higher education and to its social dimension, as identified in the strategic EU and EHEA documents discussed earlier (see the section 'Changing society through changing education'). In particular, in the Communiqué of the Leuven and Louvain-la-Neuve (EHEA, 2009), equitable access and successful completion of higher education was identified as an action line looking for the EHEA perspective of 2020: 'Access into higher education should be widened by fostering the potential of students from underrepresented groups and by providing adequate conditions for the completion of their studies' (p. 2). Participating countries are encouraged to define specific targets for the social dimension at the national level, and this process in Lithuania is led by the Research and Higher Education Monitoring and

Analysis Centre (information available at www.mosta.lt). Assessment of measures towards accessibility is integrated into 'Methodology for Evaluation of Higher Education Study Programmes' (Centre, 2010b).

Implementation of the *normativity* principle seems to be essential in seeking to balance and develop synergy between academic and political interests within higher education policy and beyond. Sustainable development always implies societal and normative choices. These, ultimately, depend on the core values of particular societies and groups which, in turn, vary between cultures across the globe and over time, defining and directing goals and framing attitudes (Waas et al., 2011). An example of this is the polarization between materialist and post-materialist values in the postmodern society which reveals a shift from an emphasis on economic and physical security towards self-expression, subjective well-being and quality-of-life concerns (Inglehart and Baker, 2000).

The normativity principle is implicitly integrated in the institutional assessment within the context indicators of the Methodology and, explicitly, as a criterion focusing on the evaluation of 'procedures to ensure adherence to academic ethics'. However, as a value-orientation principle, normativity should be included in internal QA in order to promote coherence between the HEI mission and social goals as well as to promote an understanding of academic quality per se.

Conclusion

In spite of the links between external and internal QA in Lithuanian higher education, the main principles governing internal quality framework should be slightly different to the external one in order to better facilitate transformation of the HEI towards sustainability. Specific principles for internal QA should include (1) value-orientation; (2) contextuality; (3) holistic approach; (4) reflectivity and future orientation; (5) multiple partnerships.

In implementing the methodological guidelines for institutional self-assessment and taking into account the existing regulatory framework, the following scheme is suggested for evaluating of the effectiveness and efficiency of the internal QA system as a foundation for its continuous improvement.

The effectiveness of the internal QA system shall be defined by its:

- alignment with the HEI mission and strategic goals;
- implementation of the Standards and Guidelines for Quality Assurance in the European Higher Education Area (ESG) in relation to the institution's study programmes;

- harmonization with strategic governance of the HEI and external quality assurance based on quality criteria.

The efficiency of the internal QA system shall be defined by:

- clear definition and transparency of the procedures as well as clear assignment of responsibilities for enacting it;
- regularity and the priorities set for the self-assessment procedures;
- the active involvement of all HEI staff, students and a variety of stakeholders;
- the provision of up-to-date and comprehensive information on quality procedures and results to all interested parties and society at large;
- tangible quality enhancement of the HEI's activities, including operation of the QA system, based on the self-assessment.

It is expected that, in fulfilling these criteria, QA will have an impact on moving each HEI to successfully develop and implement a fixed, 'long-term' plan using a more resilient strategic management approach that is continuously improved by an evolving QA system. Two HEIs in Lithuania – Vilnius Gediminas Technical University and Vilniaus Kolegija/University of Applied Science – have already accepted these criteria in their strategies for QA in 2013. However, the extent to which what is proposed works is a matter for future research and discussion.

Note

1. Comprehensive dynamics of the QA in Lithuania can be seen by the national reports of 2003, 2005, 2007, 2009, 2012 which are available on the website of the EHEA secretariat (EHEA, 2014).

References

Centre (2010a) 'Methodology for Conducting an Institutional Review in Higher Education' (approved by order No. 1–01–135 of the Director of the Centre for Quality Assessment in Higher Education of 25 October 2010), http://www.skvc.lt/en/content.asp?id=86, date accessed 20 April 2014.

Centre (2010b) 'Methodology for Evaluation of Higher Education Study Programmes' (approved by order No. 1–01–162 of the Director of the Centre for Quality Assessment in Higher Education of 20 December 2010), http://www.skvc.lt/en/content.asp?id=86, date accessed 20 April 2014.

EHEA (1999) 'The Bologna Declaration of 19 June 1999', joint declaration of the European Ministers of Education, http://www.ehea.info/Uploads/Declarations/BOLOGNA_DECLARATION1.pdf, date accessed 12 August 2014.

EHEA (2001) 'Towards the European Higher Education Area', communiqué of the meeting of European ministers in charge of higher education in Prague on 19 May 2001.
EHEA (2003) 'Realising the European Higher Education Area', communiqué of the conference of ministers responsible for higher education in Berlin on 19 September 2003.
EHEA (2005) 'The European Higher Education Area – Achieving the Goals', communiqué of the conference of European ministers responsible for higher education, Bergen, 19–20 May 2005.
EHEA (2007) 'Towards the European Higher Education Area: Responding to Challenges in a Globalised World', London communiqué, 18 May 2007.
EHEA (2009) 'The Bologna Process 2020 – The European Higher Education Area in the New Decade', communiqué of the conference of European ministers responsible for higher education, Leuven and Louvain-la-Neuve, 28–29 April 2009.
EHEA (2012a) 'Making the Most of our Potential: Consolidating the European Higher Education Area', Bucharest communiqué, Bucharest, 26 and 27 April 2012.
EHEA (2012b) 'The European Higher Education Area in 2012: Bologna Process Implementation Report'. EACEA P9 Eurydice.
EHEA (2014) 'EHEA Secretariat', http://www.ehea.info/article-details.aspx?ArticleId=86, date accessed 20 April 2014.
ENQA (2005) 'Standards and Guidelines for Quality Assurance in the European Higher Education Area', ENQA 4 March 2005.
EUA (2003) 'Graz Declaration 2003', http://www.eua.be, date accessed 20 April 2014.
EUA (2005) 'Developing an Internal Quality Culture in European Universities. Report on the Quality Culture Project' (Brussels: European University Association).
EUA (2006) 'Quality Culture in Universities: A Bottom-up Approach. Report on the Three Rounds of the Quality Culture Project' (Brussels: European University Association).
EUA (2009) 'Improving Quality, Enhancing Creativity: Change Processes in European Higher Education Institutions. Final Report of the Quality Assurance for the Higher Education Change Agenda (QAHECA) Project' (Brussels: European University Association).
Europa (2006) 'Delivering on the Modernisation Agenda for Universities: Education, Research and innovation'. Brussels, 10.5.2006, COM(2006) 208 final.
Europa (2010a) 'Europe 2020. A Strategy for Smart, Sustainable and Inclusive Growth'. Brussels, 3.3.2010, COM(2010) 2020 final.
Europa (2010b) 'Council Conclusions of 19 November 2010 on Education for Sustainable Development'. 2010/C 327/05.
Europa (2011) 'Supporting Growth and Jobs – An Agenda for the Modernisation of Europe's Higher Education Systems'. Brussels, 20.9.2011, COM(2011) 567 final.
Flick U. (2006) *An Introduction into Qualitative Research* (London: Sage).
Gass R. (2012) 'What is Transformation?' (Social Transformation Project), www.stproject.org, date accessed 20 April 2014.

Government (2010) 'Procedure for the External Review of Higher Education Institutions; Accreditation Procedure of Higher Education Institutions' (approved by resolution No. 1317 of the Government of the Republic of Lithuania of 22 September 2010), http://www.skvc.lt/en/content.asp?id=86, date accessed 20 April 2014.

Government (2012a) 'National Development Program for 2014–2020' (approved by decision No. 1482 of Government of the Republic of Lithuania of 28 November 2012).

Government (2012b) 'State Programme for Studies, Research and Experimental (Social, Cultural) Development for 2013–2020' (in Lithuanian) (approved by decision No. 1494 of Government of the Republic of Lithuania of 5 December 2012).

Inglehart R., Baker W. E. (2000) 'Modernization, Cultural Change, and the Persistence of Traditional Values'. *American Sociological Review*, 65, 19–51.

Loukkola T., Zhang T. (2010) *Examining Quality Culture: Part 1 – Quality Assurance Processes in Higher Education* (Brussels: European University Association).

O'Toole L. J. (2004) 'Implementation Theory and the Challenge of Sustainable Development: The Transformative Role of Learning'. In W. M. Lafferty (ed.) *Governance for Sustainable Development* (Cheltenham, UK/Northampton, MA, USA: Edward Elgar).

Seimas (2009) 'Law on Higher Education and Research' (approved by order No. XI-24230 of Seimas of the Republic of Lithuania of 9 April 2009).

Sursock A. (2011) *Examining Quality Culture: Part 2 – Processes and Tools – Participation, Ownership and Bureaucracy* (Brussels: European University Association).

Sursock A., Smidt H. (2010) *Trends 2010: A Decade of Change in European Higher Education* (Brussels: European University Association).

UNESCO (2007) *Asia-Pacific Guidelines for the Development of National ESD Indicators* (Bangkok, Thailand: UNESCO Asia and Pacific Regional Bureau for Education).

Vettori O. (2012) *Examining Quality Culture: Part 3 – From Self-Reflection to Enhancement* (Brussels: European University Association).

Waas T., Hugé J., Verbruggen A., Wright T. (2011) 'Sustainable Development: A Bird's Eye View'. *Sustainability*, 3, 1637–1661.

7
Quality System Development at the University of Graz: Lessons Learned from the Case of RCE Graz-Styria

Friedrich M. Zimmermann, Andreas Raggautz,
Kathrin Maier, Thomas Drage, Marlene Mader,
Mario Diethart and Jonas Meyer

Quality System Development at the University of Graz

Introduction and background

Owing to the development of a European Higher Education Area, the increasing autonomy of universities, and the development of global competition between universities and tertiary institutions, quality management has increasingly established itself as an organizational priority. More than ever, universities are required to make their quality system/development visible, and their achievements traceable.

At the national level, the Austrian University Act 2002 and the 2011 Act on Quality Assurance in Higher Education (HS-QSG) are in effect. The former obliges universities to take responsibility for their own quality management. The Quality Assurance Act certifies that the quality management system of universities is conducted by means of an audit with five pre-defined areas. In line with this, the University of Graz, as part of its performance agreement with the Austrian Federal Ministry of Science and Research, has incorporated a quality management system. To secure the required external evaluation of the quality management system, an audit is currently being conducted (2012–2013) by the Finnish Higher Education Evaluation Council (FINHEEC).

Quality has emerged as a leading paradigm of higher education management and of educational policy (Nickel, 2008). It not only calls for assuring the quality of education and research but also involves including focus on the needs of students, government, society and the economy in this process. Thus, depending on each specific interest group, quality management can be directed towards particular aspects of the higher education institution: excellent achievements in education and research, scientific progress, high numbers of graduates, refined periods of study and graduate employability. However, a sole focus on the targets and requirements of one interest group would of course fall short of an effective quality management system. Each can have differing expectations, different views on what 'quality' means, and this can make the focus of quality assurance and the criteria for 'excellence' highly contestable.

For this reason, discussions that centre around quality assessment in higher education are complex. Furthermore, quality management and quality assurance mechanisms not only have to be 'fit for purpose' but also have to be implemented consistently and effectively to ensure accountability. Quality assurance processes also seek to ensure that the university maintains the principles and ethical practices which protect the public and the public funding from fraud. Qualitative and quantitative indicators are used to measure the attainment of quality criteria. All this must be taken into account in quality management.

At the University of Graz the quality management system is tailored to the organization's requirements, and seeks to complement the university's internal management and structures. In order to develop a quality culture in which constructive critique and evaluation are second nature the university gives focus to the development of both the university's and its employees' competence in quality management. The quality management system is not framed as a control system but rather as a means to support the continual development of quality in education and research. We see that this can only be achieved through linking organizational learning with individual learning. In line with that objective quality management structures are designed both to serve science and society and to support a variety of methods and to align with the mission statement of the university.[1]

> As a comprehensive university, the University of Graz regards itself as an international institution for education and research committed to research and teaching for the benefit of society.
>
> It is our policy to maintain freedom in research and teaching, which permanently commits us to social, political and technological developments. Increasing flexibility and globalization are the essential frame conditions.
>
> Besides our ambition to create profile and visibility in a European and global context, it is one of our most outstanding characteristics that

our university has acquired a special position in the south-eastern European region. (Mission statement of the University of Graz)

A quality management system that is integrated and implemented into the realities of daily practice seeks constantly to develop and support the improvement of the entire university. The system of the University of Graz includes tracking and improving performance in teaching and research as well as in the management and support processes that underpin them. For every core process specific goals are set as follows:

In *the area of teaching* the University of Graz has implemented internal and external instruments (such as course evaluations, evaluations of curricula, surveys on graduates, accreditations, activities regarding the development of teaching skills and a process map for curricula development).

In *the area of research*, research evaluation is an integral component of quality management at a national as well as an international context, and it gives focus to strategic planning for the university's research activities.

At the managerial level assuring quality includes clear definition of the university's goals and ensuring that they take centre stage in operational planning and implementation.

In the *administration and service divisions* the emphasis is on consistent service quality. With this in mind, in 2013 a project for the optimization of the university's service facilities was carried out.

In order to develop and ensure quality, the autonomy, accountability and personal responsibility of all the actors has to be clear and supported. The distribution of responsibilities can be illustrated as follows:

The *Rector* oversees the quality system across the whole university, determining the realization of the system, defining the objectives, and providing the resources.

The *University Council* approves development plans and the preliminary versions of the university's performance agreement with the Government of Austria, which is then verified by the rector and rectorate.

The *Senate* issues the statutes that comprise university internal regulations such as the rules for evaluation and the rules of appointment procedures. Owing to its right to submit comments and opinions on the university's development plan, the senate is involved in this process as the plan is being shaped.

The University's Quality Management Board comprises five international experts; it counsels the university management in all matters of quality assurance and quality development as well as in questions related to quality management.

The *Office for Performance and Quality Management (LQM)* is responsible for the overall development, conceptualization and setup of the university's quality management system. LQM develops 'the basics' for quality assurance and quality development in practice, and coordinates and supports all organizational units in quality-related tasks such as the implementation of suitable tools. LQM is also responsible for preparing the system for external auditing.

The *Department for Educational and Student Services (LSS)* is responsible for the development of studies as well as for the management and development of quality management tools in learning and teaching, including the evaluation of study programmes.

The *Office of Research Management and Service (FMS)* is responsible for managing all research projects at the university level, including financial control, checking on legal aspects and intellectual property rights, overseeing projects, and reporting on all third party funding as well as on services for researchers.

Quality work in the faculties and departments is led by the *Deans* of Faculties, and the *Heads of Departments*, respectively. The deans are responsible for the arrangement of targets with the rectorate, and the academic units as well as with the staff for whom they are responsible. They are also responsible for coordinating and securing, as well as controlling, the quality of the curriculum and examinations within the departments which make up their faculty. The heads are responsible for the strategic orientation of the institute as well as conducting employee appraisals, and negotiate acquisitions and the implementation of the target agreements for the institute. This includes, inter alia, the creation and acquisition of qualification agreements and the preparation, accomplishment and implementation of research evaluation.

In addition, the *Administrative Units* are responsible for quality assurance in their respective areas of responsibility.

Students are involved in relevant ways in the formation, implementation, evaluation and accomplishment of the quality management processes. Students and *teachers* are the main players in evaluating the relevance and standard of educational content as well as the organization of teaching and studies. Students are to be informed about the quality management system through their representatives in the senate.

Purpose, key principles and goals of the quality management system

As reflected in its 2013–2018[2] strategy, the University of Graz is committed to the assurance and enhancement of the quality of its activities. Quality management is regarded as the key to achieving consistent and sustainable improvements for students, teachers and scientists, as well as for the university as a whole. Thus, the institution's quality system is embedded in its organizational processes and culture. This is seen in the university's current strategic plan:

> The range of services in teaching, research, support of early-stage researchers and service follows high-quality standards and supports continuous quality development. Thus, the resource allocation system is based on the demand arising from tasks and services on the one hand and on the profile and strategic development on the other hand, always rewarding innovative achievements. The university-wide quality culture is becoming visible by involving and integrating all university members in organizational processes. The quality management system is internationally audited every seven years (for the first time in 2012).

The purpose of the University of Graz quality system is to:

- ensure a comprehensive and systematic quality assurance and enhancement, of research, teaching, support of early-stage researchers and services;
- ensure that organizational units and their activities meet organizational objectives defined in the university's strategy;
- demonstrate compliance with internal and external requirements.

The University of Graz quality management system is based on the following key principles:

- It is aligned with the university's strategy.
- It puts emphasis on aspects of development and not on quality control.
- It is based on communication and dialogue.
- It assists teaching and research, rather than dominating them.
- It fosters a combination of clear, individual and jointly agreed target setting.
- It promotes active participation of all members of the university.
- It facilitates systematic documentation.
- It is intended to be evidence-based, and informed by data.

The quality management system at the University of Graz aims to accomplish the following goals:

- establishment of a quality culture;

- long-term implementation of university strategy;
- enhancement of the transparency of goals, processes and data;
- stronger commitment and participation of staff in quality improvement;
- incorporating quality enhancement in all areas of the university;
- set up task-oriented tools for quality management.

As stated in the development plan and the mission statement, the University of Graz sets a focus on principles of sustainability.³

> Within the scope of our social responsibility, one of our focuses is maintaining and fostering ecological, economic and social responsibility. We dedicate ourselves to issues of sustainability not only in teaching and research, but also in the field of university development. (Mission statement, development plan, p. 6)
>
> The University of Graz puts great value on a sustainable use of resources. By 2020 the CO_2 footprint will be reduced by 10% and the university set an excellent example for opportunities for a sustainable creation of infrastructure. Construction projects are realized in the most environmentally correct and ecological way. The use of energy follows the principles of climate and resource saving. (Strategic goal, development plan, p. 21)
>
> The University of Graz is committed to the principle of sustainability and the responsible use of natural resources, so that the university will be perceived as a role model for a sustainable long-term and livable society. For all substantial decisions the aspects of sustainability shall be increasingly considered in the future. (Development plan, p. 174)

Sustainability is pursued using a range of different quality tracking and improvement measures and projects. These include: (1) the introduction of a CA-FM system (computer-aided facility management) and the establishment of an appropriate energy management and climate report; (2) a pilot 'green office' project and optimizing of energy consumption (thermal and electrical); (3) the establishment of an interactive map and a barrier-free campus to enable sustainable mobility on campus, and a geographic information system (GIS) which helps to provide answers to issues of barrier-free mobility; and (4) 'sustainability in education', aiming at developing a socially sustainable university.

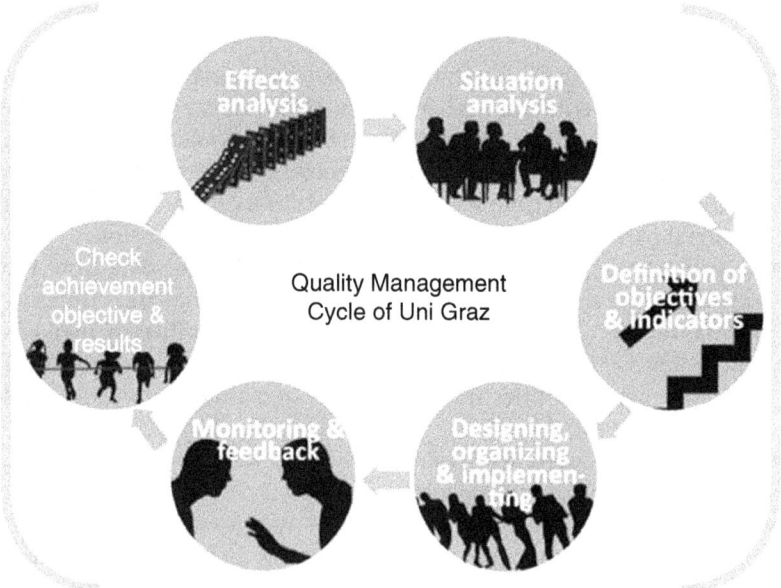

Figure 7.1 Quality management cycle of the University of Graz

Quality management cycle and key elements of the quality system

Owing to the fact that there is nothing like a standard quality management model for tertiary educational institutions and universities, the University of Graz has developed its own system, illustrated as a cycle (Figure 7.1). In developing the system, *sustainability is the basis and the result* of quality management processes. Its key elements are based upon all three pillars of sustainability (social, economic and environmental) and by using participatory approaches it is creating a dynamic process of learning, reflections and improvement.

The quality system follows the logic of the quality cycle. In simple terms it describes a cycle that runs from the establishment of objectives through planning to meet those objectives, the implementation and monitoring of actions to meet the plans, followed by the identification and analysis of outcomes and the subsequent enhancement of processes. In addition to the review on target achievement and its consequences, the effects of each initiative and its side-effects serve as a base for enhancement.

The quality system comprises a number of key elements, including:

- Continuous strategic planning processes with monitoring of operations, implementation and feedback. Performance agreements are used as a mechanism to translate the institutional objectives into expectations for faculties and staff members. The performance agreements and the extent to which the targets in them are reached are an integral part of virtually every resource allocation discussion.
- A comprehensive reporting system that provides key data from the core areas (research, teaching) and the cross-sectional areas (internationalization, gender equity, resources) assists at different management levels; it thus provides an objective basis for establishing improvement priorities for each project and their justification and communication.
- Periodic and occasional evaluations of all fields of activities (e.g. research, courses, curricula and services).
- Strategic human resource development (e.g. quality development in appointment processes, internal training and further education development on key changes including in ESD for staff, using the resources of the Competence Centre for University Teaching, leadership and organizational development). This investment in assuring the quality of staff is in recognition that they are the university's most important asset.
- Annual appraisal interviews for academic and general staff, involving goal setting and reporting on achievements and establishing possible paths for development, requirements and training.
- Tracking initiatives on students and graduates (such as the long-term career tracking of graduates, with regard to integration in the labour market, status, salary, and transition from university to professional life).
- Benchmarking initiatives[4] (e.g. in internationalization, research management, library) to gain new perspectives on performance and further development.

Quality management – a living and adapting system

In practice the implementation of a quality management system is a continual process, building upon the pre-existing measures, processes and instruments already in place. Quality measures underpin the quality cycle. Figure 7.2 illustrates the incorporation of quality measures and tools in the quality cycle and shows the relevance of having internal targets and performance agreements. It also identifies the breakdown of strategic management into operational management at different organizational levels.

Quality System Development at the University of Graz   139

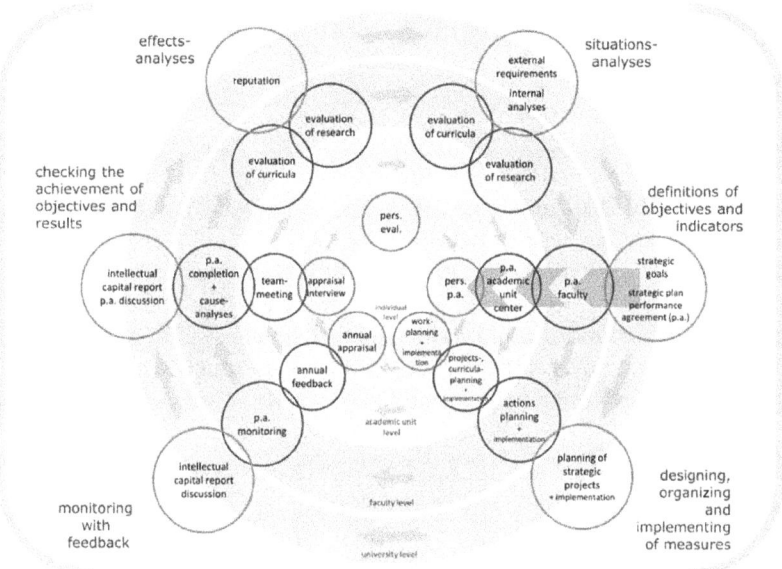

Figure 7.2 System elements of the quality management cycle at the University of Graz

The evaluative results from the quality management system are linked to strategic control by using them as a basis for target setting in performance agreements. One precondition for this process to work is the recognition, building and utilization of employees' knowledge base, imagination, experience and capacity for responsibility. This active participation can only be attained through meaningful communication and engagement, and by fostering the creative freedom and responsibility of employees in their own role.

The implementation of the quality management system significantly contributes to the success of the quality development process. An effective methodology ensures that objectives are set through the active engagement of all staff and not just of external bodies or higher management. The exchange of ideas and experience, the incorporation and usage of existing know-how and the motivation and increase of participation are all the 'soft' goals of the quality management system.

Promoting shared values and attitudes among the staff and the scientific community is more important than, for instance, the mapping of simple administrative processes. The establishment of a university community takes place by enhancing employee participation and personnel development, so that quality-related activities are perceived as a rewarding opportunity for enhancement and not as a threat. Quality improvement measures need to be aligned with university's academic programme objectives, its key values and the needs and targets of the university and its employees.

In the University of Graz approach to quality management, it is not only the achievements of its initiatives but also the effects that are systematically evaluated, using measures directly designed to measure these effects. The system helps identify both individual and institutional strengths and areas for improvement, along with links to broader innovative trends. It is in this way that the system provides the necessary framework and conditions to support quality assurance and quality development, and establish a quality culture to influence positively the university's development.

The evaluation of the RCE Graz-Styria as a case study for quality management at the University of Graz

The first evaluation using the University of Graz quality system outlined above was of a research institution, in this case of the Regional Centre of Expertise on Education for Sustainable Development, the RCE Graz-Styria. The centre already has a six-year history and the questions we faced in this case were where to start in the quality management cycle: do we start with a situation analysis and then proceed with the definition of objectives and indicators, and so on, or would it better to evaluate the achievements and the results of the RCE and, as a result, undertake an effects analysis? From our experience it was both: during the situation analysis there was a need to start with the effects analyses, using the objectives which were set by the founders of the RCE in cooperation with the dean of the faculty. This meant we started with steps five and six of the university's quality management cycle by mapping the current situation of the centre against its 'old', original objectives. The inclusion of external and internationally established researchers as evaluators in this process led to a new definition of objectives and indicators and consequently to a redesigning and revalidating of the measures used to measure implementation and impact.

The evaluation of RCE Graz-Styria is presented as a case study for quality management at the University of Graz and helps clarify the role of sustainability within quality assessment at higher education institutions. The formal evaluation process, background information about the RCE Graz-Styria, recommendations for improvement action that have emerged from the quality assessment report and findings for the development of a new strategy for the RCE, will be explained in the following sections according to the quality management cycle of the University of Graz.

Positioning the RCE internationally and within the university

The RCE Graz-Styria, acknowledged by the United Nations University (UNU) in 2007, is part of a global learning space of currently 129 RCEs worldwide (UNU-IAS, 2013). The first seven RCEs were launched in 2005. RCEs globally aim at contributing to achieve the goals of the UN Decade of Education for Sustainable Development (2005–2014) (DESD) by translating global sustainability strategies into local and regional actions (Mochizuki and Fadeeva, 2008). The DESD Monitoring and Evaluation Report (Wals, 2012) refers to the potential for RCEs to foster multi-stakeholder social learning. Universities are crucial actors in RCE networks as they undertake action-oriented research, link and leverage local sustainability projects focused on the same area, generate mutual knowledge, build capacities and competences for solving socially relevant problems, and foster transformative education for sustainable development (ESD) (Fadeeva, 2007). Universities involved in this RCE network are in the unique position 'to link the global and local on the one hand, and theory and action on the other' (Mochizuki and Fadeeva, 2008, p. 379). This is an opportunity which the University of Graz has taken up via the RCE which it hosts.

The University of Graz recognizes its societal responsibility and defines itself as a 'sustainable university'. In its mission statement (and strategy plan) the university commits itself to undertaking 'research and teaching for the benefit of society'. Addressing socially relevant questions as well as working in cooperation with (non-academic) regional actors in a transdisciplinary research setting enables the university to serve the needs of society and hence contribute actively to a regional development. Furthermore, the university's mission statement states 'we dedicate ourselves to issues of sustainability not only in teaching and research, but also in the field of university development' (University of Graz, 2013). Several initiatives over the past years have contributed to a continuous quality development process towards becoming a sustainable university. These include the production of sustainability reports, study programmes dedicated to sustainable development, ecological university management, and the establishment of institutions like the RCE Graz-Styria (cf. also Mader et al., 2011, Mader, 2013; Sedlacek, 2013; University of Graz, 2013; Zimmermann, 2006, 2007).

In 2007–2008, the RCE Graz-Styria was established in close cooperation with the Department of Geography and Regional Science as a third party funded project. In 2009 the RCE became a centre at the Faculty of Environmental and Regional Sciences and Education (URBI) at the University of Graz.

As the RCE Graz-Styria evolved in connection with the Department of Geography and Regional Science sustainable spatial development presented the main thematic focus during the first years (Mader et al., 2008). Owing to the shared vision of all RCEs to focus on education for sustainable development, especially in higher education institutions, the latter element has gained more and more importance. Following the transfer of the RCE to the Faculty of Environmental and Regional Sciences and Education the RCE Graz-Styria was positioned within two of the university's fundamental research topics, namely 'Environment and Global Change' and 'Education–Learning–Knowledge'. Furthermore, project work helped to develop specialization within the RCE, redefining its research aims from five wider focus groups of lifelong learning, research, education for sustainable development, sustainable university, and knowledge transfer (cf. Mader et al., 2008) to (1) 'Education for Sustainable Development' and (2) 'Sustainability Strategies and Sustainability Transitions'.

The process of adjusting the goals of the RCE Graz-Styria was underpinned by research projects focusing on the two research areas which were assumed to have a high potential for development within the national and international research community. Today, the RCE Graz-Styria boasts several international publications and third party projects for both research areas. In accordance with the RCE's role as an interface for innovation between university and society, publications range from the category of 'science to public' articles in books to publications in international, peer-reviewed and ranked journals (e.g. *Journal of Cleaner Production*, Elsevier; *Journal of Education for Sustainable Development*, Sage Publications; *International Journal of Sustainability in Higher Education*, Emerald).

The publications are based on the results of national and international research projects with the aim to develop transdisciplinary approaches and generate knowledge for society and university.[5]

Application of the university's quality process to RCE Graz-Styria

After the announcement that the performance of RCE Graz-Styria was to be evaluated by using the university's quality management process, the

RCE prepared a self-assessment. This is based on the criteria set down in the quality management system of the university, specifically: (1) a brief profile of the RCE (including its mission statement; development objectives and milestones; organizational structure and team; its research focuses and performance profile; its regional and international cooperation and outreach programmes; how it networks within the university; its interdisciplinary perspectives; gender issues; and the role of the RCE within a national and international context); (2) a development strategy (including a research strategy; a strategy for third party funding; a strategy to foster young researchers as well as activities for gender equity); (3) a SWOT analysis providing information about its strengths, weaknesses, opportunities and key areas for risk management.

The self-assessment helped to give structure to all of the RCE's activities and processes that had taken place between 2008 and 2012 – which was the evaluation period for the quality review. Based on this description the RCE was able to reflect on and learn from the past and, from this, to adapt its vision and reorient its focus to address future challenges. During a two-day workshop the self-assessment was presented to two external evaluators, along with the internal quality management team, the rectorate and the dean. The evaluators conducted interviews with employees of RCE Graz-Styria as well as regional and national partners and stakeholders of the RCE to test the veracity of the self-assessment. The role of the external partners was to provide feedback on regionally relevant issues and evidence supporting the quality of the RCE's sustainability transformation 'products'. Based on the results of their examination of the self-assessment and their site visits and interviews the evaluators provided their quality assessment report on good practice and their recommendations for improvement. They discussed the evaluator's quality assessment report during several internal workshops and created a new development strategy for the upcoming years based on it, which was presented to the evaluators, the quality management team, the rectorate and the dean. The final results have formed the basis for further performance agreements between the RCE and the university management.

The results of the evaluation

Monitoring and feedback

The research evaluation of the RCE Graz-Styria shows that the centre carries out a variety of projects at the national and international level. The RCE Graz-Styria is very active in the area of local, national and international networking and therefore fulfils the role of an

RCE representing the technology transfer and networking of local and regional stakeholders in the field of education for a sustainable development. Furthermore, the RCE Graz-Styria is strongly present in teaching and support of young researchers at the University of Graz. According to the research evaluation, the centre fulfils its role as an RCE and is an added value to the University of Graz. In order to effectively increase this value, a more refined focus and an institutional differentiation of management structures are advised.

This statement came at the start of the evaluation report by the two external evaluators. In addition, the evaluators provided high-quality feedback and input during their site visit at the RCE. As a consequence of the evaluation permanent monitoring instruments and engagement strategies are being established. These include monthly team meetings, and annual strategy workshops with the RCE team, the Executive Committee and the Stakeholder Board, as well as a strict monitoring of the intellectual capital report.

Designing and implementing measures

The evaluation report identified several proposals for the improvement of research and knowledge transfer of the RCE. Consistent with the quality management approach and cycle of the University of Graz, the design, organization and implementation of the follow-up measures were intensively discussed and sharpened at internal workshops. This led to a new strategy for repositioning RCE Graz-Styria which aligns with the recommendations of the evaluators. Some of the strategic issues will be discussed, following the structure of the evaluation report.

Based on the recommendation for a better and more distinct focus, future activities will concentrate on 'Education and Knowledge Creation for Sustainability Processes'. This strategic goal will be implemented by integrating sustainability research and knowledge transfer into two 'environments': (1) the university environment (cooperation with the Institute for Educational Studies, 'Teachers Education New'; contributions to the university research focus 'Education–Learning–Knowledge') and (2) the regional environment (cooperation with NGOs, the Department of Geography and Regional Science groups, cooperation with cities and regions; making contributions to the university research focus area of 'Environment and Global Change'). The adjustment of the foci of the RCE Graz-Styria is the result of a development process to align with the quality evaluation report and has involved a collaborative and participative process over a number of years.

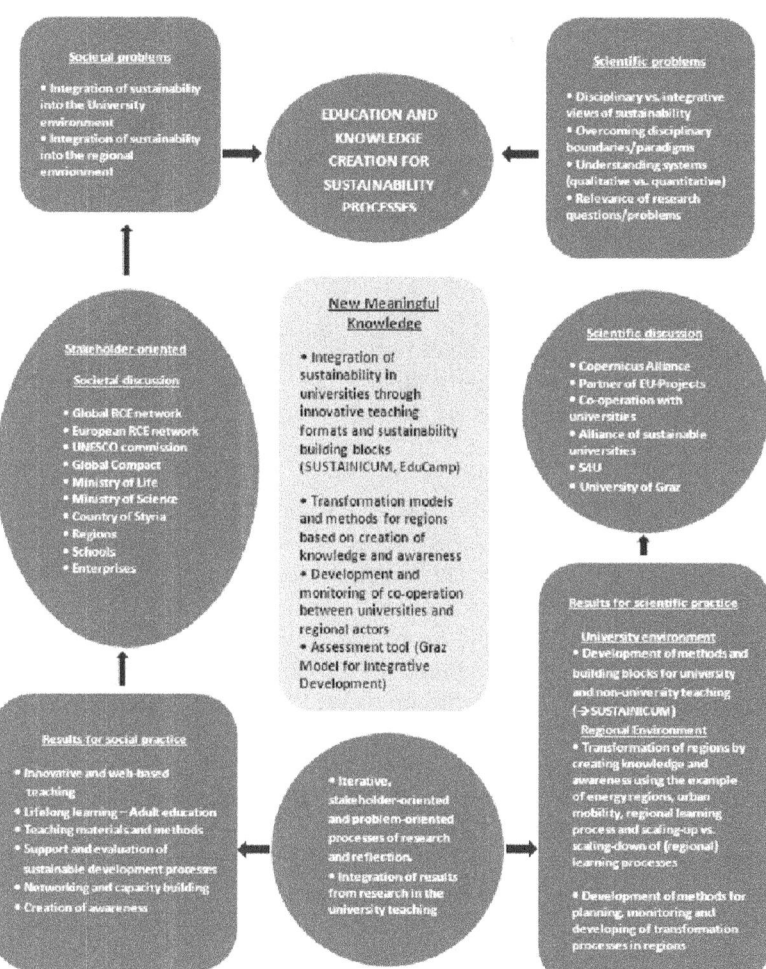

Figure 7.3 Transdisciplinary research at the RCE Graz-Styria

The detailed design and focusing of the RCE's development plan used the model of Jahn (2008) and Jahn et al. (2012): 'Transdiciplinarity in Research Practice'. An iterative and reflective process was chosen to undertake this work. Figure 7.3 shows that the RCE Graz-Styria reacts to both social problems (for example the lack of integration of sustainability in different life environments), and scientific problems, with the concept of Education and Knowledge Creation for Sustainability Processes. This model furthermore outlines the research foci, the structures and the partners of the RCE, which increases transparency, both within the university and for external stakeholders and actors. The transdisciplinary approach is also designed to contribute to the principles of the quality management at University of Graz by recognizing and utilizing the knowledge and experiences of the university staff by increasing their creativity and responsibility through participation.

Two external committees, an Executive Committee and a Stakeholder Board, are now being used to guarantee the quality and monitoring of research and knowledge transfer at the RCE Graz-Styria, supported by an internal re-organization for the provision of ongoing feedback, consultation using a clear allocation of responsibility. The establishment of an Executive Committee supports the strategic management of the RCE. This will deepen the focus on 'Education and Knowledge Creation for Sustainability Processes' at the University of Graz. The establishment of a Stakeholder Board, consisting of national and regional partners, will support the connection of the RCE Graz-Styria with the regional environment to ensure that it tackles the most relevant societal challenges. The influence of both the Executive Committee and the Stakeholder Board will contribute to transformative processes within the RCE by ensuring ongoing effects analyses and a dynamic implementation of knowledge from different actors/experts into the development process of the RCE. It is anticipated that this will also result in a higher efficiency of the RCE's work and in the optimized use of its resources.

The Executive Committee will also improve internal communication within the faculty, as well as improve targeted communication and promotion of the RCE within the University of Graz. Communication with external stakeholders along with other interest groups will take place via the RCE's digital e-mail newsletter and the 'newsletter nachhaltigkeit' (available on the RCE Graz-Styria website) which is published and printed twice a year by the Department of Geography and Regional Sciences and via oikos Graz, the student organization for sustainable economics and management.

Where to next... or... checking achievements and effect analysis

With the implementation of the new strategy of the RCE Graz-Styria, we are again at the start of a new quality improvement cycle. We have defined our strategic goals, we converted the results of the evaluation into a strategic plan and into a performance agreement with the university and we are in the process of putting into practice the different research and knowledge transfer projects identified in our new strategy. At the same time monthly team meetings, meetings with the Executive

Committee and with the Stakeholder Board will ensure effective annual appraisal and feedback consistent with the established monitoring process. This ongoing process of tracking and improvement will also be the basis for the next round of evaluation (evaluations are undertaken at five-year intervals).

It was within this context that the 'new credo' of the RCE Graz-Styria was developed:

- The basis for the Research and Development Strategy for the RCE Graz-Styria is a detailed SWOT analysis followed by external evaluation.
- Demonstrably contribute to the UN Decade of Education for Sustainable Development.
- The employees of the RCE Graz-Styria will continue to reach their ambitious goals and make a contribution to research and society in higher education via the process of 'global–local interplay'.
- The RCE Graz-Styria is to demonstrably contribute to the performance goals of the University of Graz (research focuses 'Education–Learning–Knowledge' and 'Environment and Global Change')
- Use a combination of basic, applied and transdisciplinary research to embed a focus on the societal challenges of sustainable development in the university region and in the international research community.
- Foster this approach by giving focus to the new forms of cooperation between the RCE Graz-Styria and the university, as well as societal actors and the international RCE community.
- Make 'Education and Knowledge Creation for Sustainability Processes' the core concept for the RCE Graz-Styria.
- In doing this give focus to problem-oriented action research and reflection processes, initiated by an iterative approach with key players on their needs and further developed by discussion with the scientific community.
- This structured research and development process will support young researchers, especially in identifying relevant dissertation projects for employees, and in identifying ways to integrate research results into university education.
- High-quality research and knowledge creation demands a variety of tracking measures to be effective and are vital in enacting the sustainability agenda. In addition we need to build high-quality, qualified (research-) personnel and a cooperative coordination of research activities with the faculty and university.

- Continuous monitoring and standardized communication with the Executive Committee, regional stakeholders and the university will ensure the continued relevance of this agenda and should guarantee a successful implementation of this strategy.

Conclusion

The quality evaluation and improvement processes outlined in this chapter and its direct links to the broader trends in higher education quality management, assurance and improvement processes now under way across Europe and the world has added value to both the university and the work of RCE Graz-Styria. During daily working routines it can be hard to take the time to continually reflect on the overall direction, impact and quality of research activities, structures and developments. Structured quality evaluation processes like those outlined in this chapter provide a good opportunity for all key players in a key initiative like an RCE to think over collectively and in a disciplined way what has been achieved and what next needs to be done to further improve quality. Doing this also helps with giving the RCE an evidence-based profile not only internally but internationally.

In undertaking the quality evaluation of RCE Graz-Styria we have learned that sustainability and quality management are intimately dependent on each other. Principles like reducing waste, saving energy, improving work quality, reducing disparities, improving mobility and in general identifying and improving 'defects' (one of the main goals of quality assurance) are closely connected with building a sustainable university and society. Fadeeva et al. (2012) argue that higher education institutions

> are being asked to form 'new contracts with society'... through practicing participative democracy in decision-making in all spheres of HE activities: teaching and learning, research and outreach.

Criteria at the University of Graz that provide information about the sustainability performance of a centre/department cover the role of the centre within a regional and international context, which allows conclusions regarding the engagement in regional development as well as a global exchange. The description about university-wide communication provides information about networking competences, cooperation with different departments, and a broad interest in research activities.

Strategies to foster young researchers as well as to ensure gender equity are criteria that clearly represent aspects of social sustainability. The description of the organizational structure may also provide information about the leadership of the centre, and hence the level of shared responsibilities, common visions, and participation and transparency in decision-making processes.

Criteria that could be added to enable more information about the sustainability performance of a centre include transdisciplinary aspects. For instance, to improve the society relevance of the university's outputs (which is part of the quality criteria) interdisciplinary perspectives could be extended and improved by transdisciplinary approaches. Hence the centre/department would also need to report about its research with and for the society. As higher education institutions bear a distinct responsibility to address socially and environmentally relevant challenges and have to serve societal needs, information about transdisciplinary research and education would be crucial. Activities from science to society are not that appreciated and valued so far, although lifelong learning and transdisciplinary research towards sustainable development present important aspects a university should aim at and are even stated in the mission statement of the university. Fadeeva et al. (2012) also argue that quality management has to take processes into account, such as the development of strategic and practical knowledge as well as collaborative competences. Hence a further sustainability-relevant criterion shall cover the aspect of reflection and learning. The way a centre or team reflects its research and projects provides important information about its ability in mutual learning and the generation of new knowledge. These are prerequisites for valuable sustainability research, which often deal with complex challenges that require systemic thinking and a holistic approach.

The approach to quality management at the University of Graz aims to turn the university into a learning organization and describes itself as a self-learning system that permanently needs to be adapted to new requirements. The fact that the university's mission includes the responsibility to align their research and teaching to the benefits of society and towards a sustainable development is facilitated by the process of upscaling the findings of the RCE evaluation. The discussion about evaluation criteria that particularly take sustainability issues into account is in progress and will be an innovative impulse for the next evaluation term in 2015. Consequently we can state that the work and the transformative role of the RCE is not only supporting sustainable development in the region and beyond but also encouraging the development of a sustainable University of Graz.

Notes

1. The mission statement of the University of Graz is available through the website of the university (http://www.uni-graz.at/en/university/information/about-the-university/mission-statement/).
2. The strategy of the University of Graz is set out in the current development plan for 2013–2018, available at http://static.uni-graz.at/fileadmin/Lqm/Dokumente/Entwicklungsplan_2013-2018_Uni_Graz_fuer_BMWF.pdf (only in German).
3. See development plan (http://static.uni-graz.at/fileadmin/Lqm/Dokumente/Entwicklungsplan_2013-2018_Uni_Graz_fuer_BMWF.pdf, only in German).
4. It is important to mention that the University of Graz is actively involved and is playing a key role in different international and national networks which are essential for sustainability benchmarking: for example the initiation of the network of the four Universities in Graz (Sustainability4U) and the Austrian Alliance for Sustainable Universities or the founding presidency of the COPERNICUS Alliance (European Network on Higher Education for Sustainable Development).
5. For more information about the development process of RCE Graz-Styria:

    RCE Graz-Styria was established in 2007–2008 in close cooperation with the Department of Geography and Regional Science. The founding declaration included the first definition of goals and some qualitative indicators for the RCE Graz-Styria putting an emphasis on five core areas: (1) lifelong learning; (2) research; (3) education for sustainable development; (4) sustainable university; and (5) knowledge transfer.

    The medium to long-term goals were clarified in order to establish the RCE Graz-Styria as an internationally active, competent and open institution, contributing to sustainable development on a regional, national and international level. The following aspects were defined: (1) networking with local and global sustainability initiatives; (2) consulting and project initiation in the area of Education for Sustainable Development, regional development, tourism and transfer of innovations; (3) worldwide exchange of experiences using the 'Global Learning Space for Sustainable Development', by means of the worldwide RCE network; and (4) to support awareness-raising efforts for sustainable development at the University of Graz and within society.

    The process of adjusting the goals of the RCE Graz-Styria was underpinned by research projects focusing on the two research areas which were assumed to have a high potential for development within the national and international research community: (1) 'Education for Sustainable Development' and (2) 'Sustainability Strategies and Sustainability Transitions'. Today, the RCE Graz-Styria boasts several international publications and third party projects for both research areas. In accordance with the RCE's role as an interface for innovation between university and society, publications range from the category of 'science to public' (for example UNIZEIT, Issue 01/2012), articles in books (for example GUNI Reports: *Higher Education in the World: Handbook of Good Practices of Multi-Actor Learning*) to publications in international, peer-reviewed and ranked journals (for example *Journal of Cleaner Production*, Elsevier; *Journal of Education for Sustainable Development*,

Sage Publications; *International Journal of Sustainability in Higher Education*, Emerald).

The publications are based on the results of national and international research projects with the aim to develop transdisciplinary approaches and generate knowledge for society and university. The starting point was a research project funded by the National Bank of Austria on the topic 'Transition Processes in Regions' (PhD thesis by C. Mader, supervised by F. M. Zimmermann, Mader C., 2009). The project resulted in the Graz Model for Integrative Development which was developed into an assessment tool for development processes in regions, businesses and organizations. The VCSE Project ('Virtual Campus for a Sustainable Europe', EU e-learning programme 2007–2009) focused on sustainability in higher education. Several online teaching sessions about sustainability topics were developed and are still offered by partner universities. As part of the VCSE Project, the COPERNICUS Alliance (the European Network on Higher Education for Sustainable Development) was founded. The follow-up project was the 3-LENSUS ('Lifelong Learning Network for Sustainable Development', EU Lifelong Learning Programme (LLP), 2009–2010) which focused on the establishment of the structure of the COPERNICUS Alliance (and hosted the founding presidency by F. M. Zimmermann at the University of Graz). Furthermore, innovative case studies about sustainability were developed, connecting university research with local actors. Another EU LLP project (2010–2012) is the ICTeESD (ICT enabled Education for Sustainable Development) which was dedicated to the development of a joint master's degree with ICT supporting education for sustainable development. This strengthened the e-teaching competences of the RCE Graz-Styria. Within the EduCamp project ('Education for Sustainable Development beyond the campus', EU Tempus, 2010–2013), education centres at seven Egyptian universities were founded and innovative teaching materials about sustainability for professors and students were developed. The trained students and professors will educate teachers in Egypt, generating a multiplier effect. The national 'Sustainicum' project (funded by the Austrian Federal Ministry for Science and Research, 2012–2013) has the aim to develop learning blocks and teaching methods about sustainability topics and to make them publicly available for higher education purposes on the internet.

References

Fadeeva Z. (2007) 'From Centre of Excellence to Centre of Expertise: Regional Centres of Expertise on Education for Sustainable Development'. In A. E. J. Wals (ed.) *Social Learning Towards a Sustainable World* (Wageningen: Wageningen Academic Publishers), 245–264.

Fadeeva Z., Galkute L., Lotz-Sisitka H., Razak D. A., Chacón M., Yarime M., Mohamedbhai G. (2012) 'University Appraisal for Diversity, Innovation and Change Towards Sustainable Development? Can it be Done?' In Global University Network for Innovation (ed.) *Higher Education in the World 4: Higher Education's Commitment to Sustainability: from Understanding to Action* (New York: Palgrave Macmillan), 310–315.

Jahn T. (2008) 'Transdisziplinarität in der Forschungspraxis'. In M. Bergmann, E. Schramm (eds) *Transdisziplinäre Forschung. Integrative Forschungsprozesse verstehen und bewerten.* (Frankfurt/New York: Campus Verlag), 21–37.

Jahn T., Bergmann M., Keil F. (2012) 'Transdisciplinarity: Between Mainstreaming and Marginalization'. *Ecological Economics*, 79, 1–10.

Mader C. (2009) *Principles for Integrative Development Processes towards Sustainability in Regions – Cases Assessed from Egypt, Sweden and USA* (Graz: University of Graz).

Mader C. (2013) 'Sustainability Process Assessment on Transformative Potentials: The Graz Model for Integrative Development'. *Journal of Cleaner Production*, 49, 54–63, http://dx.doi.org/10.1016/j.jclepro.2012.08.028.

Mader C., Mader M., Diethart M. (2011) 'Der Nachhaltigkeitsprozess der Universität Graz – analysiert durch das Grazer Modell für Integrative Entwicklung'. In F. M. Zimmermann (ed.) *Nachhaltigkeit, Regionalentwicklung, Tourismus, Festschrift zum 60. Geburtstag von Friedrich M. Zimmermann* (Graz: Grazer Schriften der Geographie und Raumforschung), 46, 63–70.

Mader C., Zimmermann F. M., Steiner G., Risopoulos F. (2008) 'Regional Centre of Expertise (RCE) Graz-Styria – A Process of Mobilization Facing Regional Challenges'. *International Journal of Sustainability in Higher Education*, 9 (4) 402–415.

Mader M., Mader C., Zimmermann F. M., Görsdorf-Lechevin E., Diethart M. (2013) 'Monitoring Networking between Higher Education Institutions and Regional Actors'. *Journal of Cleaner Production*, 49, 105–113.

Mochizuki Y., Fadeeva Z. (2008) 'Regional Centres of Expertise on Education for Sustainable Development (RCEs): An Overview'. *International Journal of Sustainability in Higher Education*, 9 (4) 369–381.

Nickel S. (2008) 'Qualitätsmanagementsysteme an Universitäten und Fachhochschule: Ein kritischer Überblick'. *Beiträge zur Hochschulforschung*, 30 (1) 16–39.

Sedlacek S. (2013) 'The Role of Universities in Fostering Sustainable Development at the Regional Level'. *Journal of Cleaner Production*, 48, 74–84.

University of Graz (2013) 'Development Plan 2013–2018', http://static.uni-graz.at/fileadmin/Lqm/Dokumente/Entwicklungsplan_2013-2018_Uni_Graz_fuer_BMWF.pdf (only in German), date accessed July 2013.

UNU-IAS (United Nations University – Institute for the Advanced Study of Sustainability) (2014) 'Education for Sustainable Development – RCEs and ProSPER.Net', http://ias.unu.edu/en/research/education-for-sustainable-development-rces-and-prosper-net.html#outline, date accessed July 2014.

Wals A. E. J. (2012) 'Shaping the Education of Tomorrow: 2012 Full-length Report on the UN Decade of Education for Sustainable Development', UNESCO DESD Monitoring and Evaluation (Paris: UNESCO).

Zimmermann F. M. (2006) 'Nachhaltige Entwicklung und Universitäten'. In G. Steiner, A. Posch (eds) *Innovative Forschung und Lehre für eine nachhaltige Entwicklung, Bericht aus den Umweltwissenschaften* (Aachen: Shaker), 1–13.

Zimmermann F. M. (2007) 'The Chain of Sustainability'. In PSCA International (ed.) *Public Service Review: European Union* (Newcastle under Lyme, UK: PSCA International Ltd.Publication), 232–233.

8
STARS as a Multi-Purpose Tool for Advancing Campus Sustainability in US

Monika Urbanski and Paul Rowland

The Role of STARS in the campus sustainability movement

Emergence and growth of sustainability in higher education

Over the past two decades, the emergence and growth of campus sustainability has contributed to a broad global movement calling for a transition toward a socially just and sustainable future for all. 'Higher education institutions bear a profound, moral responsibility to increase the awareness, knowledge, skills, and values needed to create a just and sustainable future' (Cortese, 2003). In addition to preparing the leaders of tomorrow, higher education serves as a sustainability model through its influences on K-12 learning and its role as a source for new ideas, commentary on social challenges, and engaged experimentation in sustainable living.

The Association for the Advancement of Sustainability in Higher Education (AASHE) is one of several organizations in North America that was established as a result of the growing campus sustainability movement. AASHE provides thought leadership, resources, professional development opportunities and a framework for demonstrating the value and competitive edge created by sustainability initiatives.

The need for tracking sustainability performance

As colleges and universities began including sustainability as a key component within mission and strategic direction, there has been a growing realization of the significant challenges to its successful application. Some of these challenges arose from a historical lack of easily accessible guidance on the subject: 'The literature on sustainability in higher education (and related fields) provides little systematic guidance to potential change agents about the process of becoming a sustainable campus' (Shriberg, 2002, p. 86). In addition, lack of data on sustainability metrics has historically made it difficult for administrations to establish goals and measure progress: 'While extensive information is available on most other aspects of performance – enrollments, costs, state regulations, competitiveness, and demographics, for example, no such source on environmental performance exists...consequently, there has been no baseline from which to measure progress across a range of issues' (McIntosh et al., 2001). It was in response to this need that the Sustainability Tracking, Assessment & Rating System (STARS) was developed.

History of STARS

A collaborative effort to address the growing need for standard sustainability metrics led to the development and release of the Sustainability Tracking, Assessment & Rating System (STARS). In 2006, the Higher Education Associations Sustainability Consortium (HEASC) issued a call for the development of a standardized campus sustainability assessment tool. HEASC envisioned a comprehensive system that would 'address all the dimensions of sustainability (health, social, economic and ecological) and all the sectors and functions of a university, including curriculum, facilities, operations, and collaboration with communities'. Soon after AASHE was formed in 2006, HEASC called upon AASHE to 'convene all relevant stakeholders in a collaborative process to develop such a system' (HEASC, 2006). AASHE responded by leading the development of STARS through a collaborative, three-year process.

Box 8.1 About AASHE and STARS

The Association for the Advancement of Sustainability in Higher Education (AASHE) is a non-government organization that promotes sustainability in higher education in the US and Canada, with several internationally focused projects and initiatives. With more than 1100 institutional members, AASHE provides high-quality resources, professional development opportunities, and thought leadership on sustainability as a critical part of campus operations and curriculum. A programme of AASHE, the Sustainability Tracking, Assessment & Rating System (STARS), is a transparent, self-reporting framework that allows colleges and universities to measure their sustainability performance. With STARS, institutions can assess sustainability performance in a number of areas including curriculum and research, student and public engagement, campus operations, investment and administration.

Launched in 2009, STARS is intended to engage and recognize a full spectrum of colleges and universities in their sustainability efforts – from community colleges to research universities, and from institutions just starting their sustainability programmes to long-time campus sustainability leaders. Although STARS was designed for US and Canadian institutions, growing interest at the international level resulted in the launch of the STARS International Pilot in 2011. The pilot allowed colleges and universities outside of Canada and the US to publicly document their sustainability efforts, with opportunities for these institutions to share feedback and make suggestions for improvements to the system.

STARS participation has grown steadily since 2009, surpassing 300 rated institutions in 2014. Institutions in Mexico, Japan, Pakistan and France were among the first to submit a report under the STARS International Pilot. The development of STARS 2.0 was a key focus in 2012 and 2013. Released in late 2013, STARS 2.0 facilitates 'meaningful assessments of campus sustainability performance while remaining accessible and relevant to the diversity of higher education institutions' (AASHE, 2013, p. 5). The release of STARS 2.0 is expected to improve applicability of STARS to higher education institutions located outside of the US and Canada.

Key components of STARS

STARS is a voluntary, self-reporting assessment that allows institutions to accumulate points for various campus sustainability activities within academic programmes, operations and administration. The STARS 2.0 system[1] measures sustainability performance in five categories: Academics (AC), Engagement (EN), Operations (OP), Planning & Administration (PA), and Innovation (IN). By participating in STARS, institutions gain access to a framework for understanding sustainability in all sectors of higher education, as well as to meaningful comparisons over time and across institutions using a common set of measurements. Use of the tool creates incentives for continual improvement toward sustainability and facilitates information sharing about higher education sustainability practices and performance. Collectively, STARS participation helps to build a stronger, more diverse campus sustainability community.

Participating institutions submit their evaluation materials online through a reporting tool and can be awarded a rating ranging from Bronze to Platinum based on cumulative points earned. Institutions that do not seek a rating can receive Reporter status and can still take advantage of the tracking and assessment features of the reporting tool. STARS is subscription-based, providing institutions with 12 months of continuous access to the reporting tool. A STARS discount is available for AASHE member institutions.

The Role of STARS in global higher education sustainability assessment

While several sustainability assessment tools have been developed worldwide as a result of growing demand in the higher education community,

STARS has emerged as the premier assessment for comprehensive sustainability performance among North American higher education institutions.

Other North American organizations collect and disseminate data on specific aspects of sustainability performance. For example, institutions may voluntarily submit data related to climate and greenhouse gas emissions to the American College and University Presidents Climate Commitment (ACUPCC). Similarly, institutions may pursue Leadership in Energy and Environmental Design (LEED) ratings for campus new construction and renovations. While many of these organizations focus on specific components of sustainability, STARS on the other hand provides institutions with the opportunity to report on comprehensive campus sustainability outcomes. Included in the STARS reporting tool are metrics for building design and maintenance, energy and climate, water use and waste diversion, food, purchasing, diversity and affordability, human resources, research, and curriculum among other topics. While there are a few other comprehensive campus sustainability rating programmes for North American institutions, these are not accessible to all institutions (for example, two-year institutions may be excluded).

The comprehensive nature of STARS has enabled collaboration among North American organizations in efforts to streamline the reporting process for institutions that choose to participate in various higher education sustainability assessments. In 2011, The Princeton Review, Sierra Magazine and the Sustainable Endowments Institute (SEI) worked with AASHE to establish the Campus Sustainability Data Collector as a tool for sharing sustainability data. The result of this collaboration allowed institutions to share data snapshots through the STARS reporting tool platform free of charge. The data may then be shared with Sierra Magazine and The Princeton Review for their annual sustainability publications (SEI suspended production of the College Sustainability Report Card in 2011 and is no longer seeking data for this publication). With the release of STARS 2.0 in 2014, access to STARS was further enhanced with the introduction of STARS Basic and Full Access. Available at no cost, Basic Access allows institutions to use the STARS Reporting Tool to track and share data with other organizations. Full Access to STARS goes a step further, allowing institutions the option to submit for a STARS rating (Bronze, Silver, Gold, or Platinum). By 2014, over 650 higher education institutions have participated in STARS through Full or Basic Access, resulting in over 700 data snapshots shared.

Sustainability assessment tools have been developed beyond North America to address the universal need for sustainability metrics in higher education worldwide. While AASHE has not yet developed formal partnerships with these international organizations, informal collaboration and information sharing is occurring on an ongoing basis. At the AASHE Annual Conference in 2012 for example, a workshop entitled Measuring Campus Sustainability around the World provided attendees an overview of five higher education sustainability assessment tools. Among these assessments are STARS (US and Canada), Learning in Future Environments (LiFE) (UK and Australasia), The Green Plan (France), Auditing Instrument for Sustainability in Higher Education (AISHE) (Netherlands and Sweden) and Alternative University Appraisal (AUA) (developed by ProSPER.Net in Japan) (AASHE, 2012a).

Limitations and challenges: STARS and sustainability assessment overall

While the benefits of using STARS are clearly outlined throughout this chapter, it is equally important to note some of its limitations, as well as limitations for higher education sustainability assessments in general. Regardless of what type of assessment tool is employed, the diverse nature of higher education institutions makes comparisons and benchmarking universally challenging. This challenge is particularly relevant for STARS owing to its comprehensive nature and its intent of being applicable to all institution types and contexts.

The challenges related to institutional diversity are perhaps most evident in the international context. Through feedback from institutions participating in the STARS International Pilot and at the 2012 international workshop mentioned previously, we have learned that several areas within STARS are not ideally suited to institutions outside of the US and Canada. The following constraints have been identified by some international institutions participating in STARS (a number of these concerns have been addressed in changes included in the release of STARS 2.0):

- not all cultures rely heavily on quantitative measures of assessment;
- reliance on LEED certification and ratings within the Buildings subcategory (LEED is predominantly a North American rating system);
- laws and policies that limit freedom in purchasing and procurement;
- units of measurement applied in STARS that may not be universally applicable, requiring added learning and unit conversion;
- limitations on collecting certain types of demographic data, particularly statistics on human resources and student and staff diversity and accessibility.

To a lesser extent, the issue of institutional diversity has been found to pose challenges among North American institutions. As outlined later in this chapter, differences in institution type and other factors are found to significantly impact sustainability performance. Adding to the challenge of institutional diversity, climate attributes represent an additional set of variables that can significantly impact on sustainability performance for higher education institutions. Climate has a major impact on building design, construction and energy use, resulting in differing energy code standards by region. Variations in temperature, moisture and precipitation, elevation, and even recent weather patterns have the potential to differentiate institutions of higher education that may otherwise be very similar. The higher education field currently lacks benchmarking methods that incorporate climate variations and institutional diversity (Jaye, 2012). In attempting to classify institutions, consideration of environmental factors is important, but may result in complications in defining peer institution groups.[2]

To address these challenges, AASHE is continuously working toward clarifying how institutional diversity and climate impact sustainability performance, and building into the STARS reporting process clear and relevant data filters to allow for easier benchmarking. In doing so, AASHE is introducing periodic feature upgrades to the STARS data displays. In 2013, for example, AASHE updated its Organization Type filter to allow for differentiation among research, non-research, and two- and four-year institutions. These changes allowed for more detailed benchmarking and offered greater opportunities for learning about how institution type impacts sustainability performance.

Because of the self-reporting nature of STARS, data integrity is another significant challenge, probably one that applies to other sustainability assessment tools as well. To promote STARS data accuracy, all data submitted by participating institutions become public once reports are submitted. If an individual or organization believes that erroneous data have been submitted, a STARS submission accuracy enquiry can be submitted, anonymously if desired, to bring the potential error to the attention of the STARS liaison at that institution. These enquiries can then be addressed through data correction requests submitted by STARS liaisons. Although STARS data are not verified by AASHE or a third party, AASHE staff members conduct random reviews of credits within each submission as well as periodic audits of comparative data. These random reviews and periodic audits have resulted in numerous approved data corrections. AASHE is working to continuously improve processes for ensuring data quality, with several new data accuracy measures introduced alongside the release of STARS 2.0.

Leading the sustainability transformation through STARS

The remainder of this chapter focuses on five specific ways in which sustainability performance at higher education institutions can be enhanced through STARS. Insights gained from STARS data analysis, as well as feedback obtained from AASHE annual conference attendees, STARS webinar participants, and the STARS post-submission survey tell a compelling story of how STARS is facilitating the transition toward a more just and sustainable society.

Through its impact on campus operations and social dimensions, STARS fulfils a transformational role in advancing campus sustainability. STARS helps to transform campus operations by helping institutions to: (1) conduct a gap analysis of sustainability performance, and (2) benchmark sustainability performance and make comparisons with peers. STARS helps to transform social dimensions related to sustainability by: (3) starting campus conversations to advance sustainability, and (4) engaging students in campus sustainability initiatives. STARS combines social dimensions with campus operations to promote the sustainability transformation through a fifth function: (5) using STARS data to learn about the current state of campus sustainability.

Box 8.2 Five methods for transforming campus sustainability through STARS

- *Conducting a gap analysis of sustainability performance:* Institutions can improve campus operations by completing a STARS report, earning a rating, and conducting a gap analysis through review of category, subcategory and credit scores.
- *Benchmarking sustainability performance for comparison with peers:* Synergistic qualities of benchmarking can result in sustainability advancements that have the potential to transform the global sustainability movement.
- *Starting conversations to advance sustainability:* Through social interactions, institutions can garner support and buy-in to help transform campus sustainability. Participation in STARS encourages campus dialogue on defining sustainability and identifying areas for improvement.
- *STARS as a tool for engaging students:* Engaging students in STARS offers a number of benefits to students and also for the institution; however, the greatest benefit is the potential that student engagement has on advancing sustainability on a broad scale.
- *Using STARS data to learn about the current state of campus sustainability:* A wealth of information on trends and good practices is emerging as a result of the public nature of STARS data submissions. By communicating these findings broadly, advancements in campus sustainability can be achieved more quickly.

Conducting a gap analysis of sustainability performance

Perhaps the most widely recognized function of STARS is its role in providing accurate, comparable information about institutions' actual sustainability performance and potential for improvement. Completing STARS helps to answer the questions: 'How are we doing in addressing sustainability on our campus?' and 'Where do we want to be, in terms of our sustainability performance?' STARS provides numerous opportunities to measure current sustainability performance and plan for improvement. Institutions can improve campus operations by completing a STARS report, earning a rating, and conducting a gap analysis through review of category, subcategory and credit scores. Collectively, this function has a significant impact on promoting sustainable operations that lead to a broad sustainability transformation.

At the AASHE 2012 Annual Conference, a Babson College sustainability staff member outlined how the college used STARS to identify strengths and weaknesses in its sustainability performance (Scott, 2012). After submitting a follow-up report a year later, sustainability performance in a number of STARS subcategories improved. Through its earned rating and review of category, subcategory and credit scores, the college was able to identify several short-term areas for improvement.

Earning a STARS rating

Institutions that wish to be scored may earn one of four STARS ratings: Bronze, Silver, Gold and Platinum. Participating institutions that do not want to publish their scores may participate as a STARS Reporter. As of 2014, no institution had achieved the highest rating of STARS Platinum. Fifty-one per cent of institutions earned a Silver rating, 22 per cent earned a Bronze rating, and 20 per cent earned a Gold rating. Seven per cent of institutions submitted a report under a STARS Reporter designation (including four International Pilot participants).

Institutions can track sustainability progress over time by regularly submitting STARS reports. Some institutions, particularly those that have been early adopters of campus sustainability, have set ambitious goals of leading the higher education sustainability movement. For such institutions, earning a STARS Gold or even Platinum rating may be a viable goal-setting strategy. Other institutions are primarily interested in tracking performance and setting gradual goals for overall or targeted improvement. Performance in certain categories, subcategories and credits may be a more important strategic focus for such institutions.

Table 8.1 STARS ratings summary

Rating	% of participating institutions
Platinum	–
Gold	20%
Silver	51%
Bronze	22%
Reporter	7%

Source: AASHE, 2014.

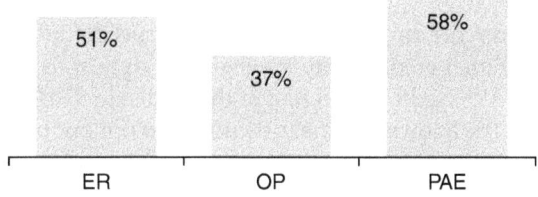

Figure 8.1 2014 analysis of average scores for all institutions under STARS Version 1
Source: AASHE, 2014.

Category scores

To gain an understanding of sustainability strengths and areas for improvement, institutions that have completed a STARS report can compare STARS categories scores. When conducting a gap analysis, it is important to note that performance varies significantly across categories. In STARS Version 1 reports, average points earned are typically highest in the PAE category and lowest in the OP category.

Subcategory scores

Every completed STARS report publicly displays an institution's scores in each of 17 STARS subcategories (although categories were reorganized with the release of Version 2.0, the STARS subcategories remained largely unchanged). As with categories, subcategory scores differ significantly owing to the nature of the subject matter within each subcategory. Analysis of STARS Version 1 score data in 2013 found that scores were highest in the Diversity & Affordability subcategory within PAE. Institutions also earned high scores in Coordination & Planning, Co-curricular Education, Grounds and Human Resources. Lowest scores were found in the Investment category, also in PAE. Other low-scoring subcategories included Energy, Climate, Buildings and Transportation, all found in the OP category.

Table 8.2 Five highest- and lowest-scoring STARS subcategories

Highest-scoring	Lowest-scoring
Diversity & Affordability – PAE	Investment – PAE
Coordination & Planning – PAE	Energy – OP
Co-Curricular Education – ER	Climate – OP
Grounds – OP	Buildings – OP
Human Resources – PAE	Transportation – OP

Source: AASHE, 2013.

Credit status and scoring

Institutions have the opportunity to address specific areas of sustainability as individual credits, and are given the option to pursue or not pursue each STARS credit. About half of the credits in STARS also include applicability rules, allowing some institutions to opt out of credits based on certain qualifying characteristics. For example, institutions that do not include research in faculty tenure and promotion decisions may select 'Not Applicable' for credits within the Research subcategory. Likewise, institutions that do not have on-campus dining services may opt out of OP 6: Food & Beverage Purchasing.[3]

Once data are finalized and submitted in STARS, gaps in performance can be identified by looking at credits that were not pursued or garnered low points from the credit total. As with STARS categories and subcategories, there is significant variation in scoring among STARS credits. In 2013, all high-scoring credits originated within PAE with the exception of Hazardous Waste Management, found in OP. Credits for renewable energy and sustainable investment earned the lowest scores in STARS, at 3 per cent and 6 per cent of total applicable points earned, respectively.

Using STARS results to perform gap analysis has a significant potential for advancing sustainability at participating institutions. As more and more institutions participate in STARS, the collective impact for the broader sustainability movement will continue to grow.

Benchmarking sustainability performance

Benchmarking is an essential tool that allows institutions to identify sustainability leadership examples for various types of institutions. It offers opportunities to learn of good practices employed by sustainability peer institutions as well as from aspirant sustainability leaders. By using STARS as a benchmarking and comparison tool, continuous and sometimes rapid progress in sustainability performance can be achieved. Benchmarking through STARS helps to answer the question: 'How are

Table 8.3 Five highest- and lowest-scoring STARS credits

Highest-scoring	Lowest-scoring
PAE 1: Sustainability Coordination	OP 8: Clean & Renewable Energy
PAE 19: Community Sustainability Partnerships	PAE 18: Positive Sustainable Investments
OP 21: Hazardous Waste Management	OP 5: Greenhouse Gas Emissions Reduction
PAE 10: Affordability & Access Programmes	OP 1: Building Operations & Maintenance
PAE 8: Support for Underrepresented Groups	ER 13: Sustainability Literacy Assessment

Source: AASHE, 2013.

we doing in sustainability compared to other institutions?' Answers to this question can most frequently be found by using the STARS data displays, which provide summaries of STARS score and content data across all institutions or based on certain institutional characteristics. Visitors to the STARS website can use the data displays to filter on characteristics such as institution type, country of origin and enrolment size. Another resource for accessing data is the STARS application programming interface (API), which allows remote read and write access to the STARS reporting system in an automatic and programmable way.

Benchmarking can benefit individual institutions by providing a competitive edge within an uncertain economic climate. Institutions worldwide are challenged by any combination of factors, including shrinking budgets, growing costs, difficulties in meeting enrolment projections, the emergence of new competitors, and the perceived value of a college or university diploma. Coupled with these constraints, there is evidence suggesting that 'how an institution is doing in sustainability' is a deciding factor for students and their parents. Results from the 2013 College Hopes and Worries Survey Report found that 62 per cent of respondents overall said that comparative information on an institution's commitment to environmental issues would contribute 'strongly', 'very much' or 'somewhat' to college selection decisions (The Princeton Review, 2013). By advancing sustainability and striving toward sustainability leadership, institutions can achieve greater marketability and may ultimately earn a competitive edge.

STARS is meant to create incentives for continual improvement rather than to rank institutions based on sustainability performance. Nonetheless, rivalry and competition exist, and are ingrained in human nature. A strong argument can be made for the use of ratings and

Table 8.4 Impact of environmental issues on college admissions

'If you (your child) had a way to compare colleges based on their commitment to environmental issues, how much would this contribute to your (your child's) decision to apply to or attend a school?'

22%	Strongly or Very much
40%	Somewhat
29%	Not much
9%	Not at all

Source: The Princeton Review, 2013.

rankings to drive sustainability performance collectively and for individual institutions. Conducting benchmark analysis of STARS score data can deliver data that may encourage the spirit of competition and a sense of excitement that can have noticeable results on individual and collective sustainability performance.

Creative and collaborative benchmarking efforts have the potential not only to create a competitive edge, but also to transform the scope and pace of sustainability advancement in society as a whole. When well conducted, benchmarking may result in synergies of effort. A recent presentation at a State University of New York (SUNY) sustainability conference demonstrated how this can occur within a university system (Dautremont-Smith, 2012). A benchmark analysis was conducted for all SUNY system institutions participating in STARS, identifying areas of strength and opportunities for improvement for all participating institutions. By encouraging the use of STARS across an entire university system, potential for system-wide advancements in sustainability is increased through opportunities for collaborative competition. Synergistic qualities of benchmarking can result in sustainability advancements that have the potential to transform the global sustainability movement.

Benchmarking can be applied at various scales to achieve targeted results. Below are some examples of how benchmarking can be applied creatively to further institutions' sustainability initiatives:

- Institutions that have earned a STARS rating can review data of other rated institutions, with a special focus on credits where there is desire or potential for improvement.
- Institutions actively pursuing a STARS Gold or higher rating may apply the STARS Rating filter to collect best practices from institutions that have earned a Gold rating.

- A group of universities participating in a sustainability colloquium can encourage use of STARS across the colloquium to facilitate information sharing.
- A community college wishing to identify networking opportunities with other two-year colleges may apply the Associate Organization Type filter to generate a list of two-year institutions that have earned a STARS rating.
- A university wishing to establish a friendly sustainability competition within its National Collegiate Athletic Association (NCAA) conference rivals can scan the Rated Institution list to identify NCAA peers that have participated in STARS.

Conversations to advance sustainability

STARS has been successful in expanding campus understanding of sustainability and has been a conversation driver and connector. Participation in STARS encourages social interactions around sustainability as a topic, which provides opportunities to educate campus community members about the benefits of sustainability. Garnering support and buy-in through various levels of communication is important in helping to transform campus sustainability.

Participation in STARS helps to answer the question, 'How can we engage and communicate with campus stakeholders to advance sustainability on our campus?' At AASHE conferences and STARS webinars, AASHE staff members hear stories about the many new conversations that are taking place and new connections being made as a result of STARS participation. The human resources staff at one university were thrilled to learn that the work they were doing counted as part of the campus sustainability effort. Another institution reported that despite some trepidation about approaching faculty to collect information on sustainability in the curriculum, they instead encountered support and excitement about this portion of the assessment.

While the act of signing up for STARS signifies a commitment of intent for improving sustainability performance, the communication processes related to completing a STARS report may play an even greater role in advancing campus sustainability. Because of its comprehensive nature, STARS requires expertise from diverse administrative and staff personnel, faculty representatives and campus sustainability advocates. The STARS reporting process has numerous opportunities for communication built in. At minimum, STARS creates communication pathways between the STARS liaison, executive contact, responsible parties assigned to credits, and the institution's highest ranking administrator, who submits a letter affirming the accuracy of the STARS report.

Through communication with STARS liaisons and website users, three areas have emerged as common conversation topics. These include: (1) defining sustainability in the curriculum; (2) incorporating the three dimensions of sustainability into STARS reports; and (3) identifying areas for improvement after submitting a STARS report.

Defining sustainability in the curriculum

For *ER 5: Sustainability Course Identification* under STARS Version 1, institutions are asked to develop a definition of sustainability in the curriculum and use this definition to develop an inventory of sustainability-focused and sustainability-related courses. While methodologies for developing a definition of sustainability in the curriculum vary greatly among institutions, a qualitative analysis of STARS data in 2013 uncovered several dominant strategies for defining sustainability in the curriculum and creating an inventory of sustainability-focused and sustainability-related courses. Analysis of the field, 'A brief description of the methodology the institution followed to complete the inventory' showed that review of course content was the primary methodology for developing a sustainability course inventory. For many institutions however, this was just one step in a more detailed process. Forty-eight per cent of institutions that reviewed course content also included other strategies such as communication with faculty and department chairs, surveys, meetings or interviews. The qualitative analysis also yielded the following results:

- 28 per cent of institutions collected input from faculty and/or department chairs through surveys or related means
- 27 per cent of institutions formed new committees/taskforces (or charged existing ones) to address sustainability course content
- 14 per cent of institutions indicated they worked closely with department/college representatives
- 14 per cent of institutions sought faculty input (other than through a survey)
- 9 per cent of institutions involved student workers, volunteers or student groups/clubs
- 7 per cent of institutions indicated they worked closely with other offices, such as the Registrar or Institutional Research
- 5 per cent of institutions indicated that courses defined as sustainability-focused or sustainability-related were promoted through various means across the institution

- 5 per cent of institutions created interactive websites or databases as a means for communicating with faculty about sustainability course content
- 4 per cent of institutions indicated that significant outreach was made to address sustainability in the curriculum through one-on-one interviews, town-hall meetings, and/or retreats. (AASHE, 2013)

Incorporating the three dimensions of sustainability in campus planning

The concept of 'sustainability' is viewed and interpreted differently by higher education stakeholders. While some interpret sustainability as a concept that is focused on environmental issues, others adopt a more holistic view. In the quest for creating an open and sustainable society, institutions must look beyond the traditional 'green' line of thinking and focus on the relationship between social, economic and environmental dimensions of sustainability.

Analysis of STARS data has found that institutions commonly include data related to environmental dimensions of sustainability in their STARS reports. Social dimensions are occasionally highlighted, while economic dimensions are very often excluded. Qualitative analysis of STARS content in 2012 found that the inclusion of the three dimensions of sustainability in STARS reports may in part be explained by language used in the STARS Technical Manual. Institutions were more likely to highlight all three dimensions if such a recommendation was explicitly made in the manual. For credits where the manual did not explicitly encourage inclusion of social and economic dimensions, there was a tendency for institutions to report predominantly green behaviours (AASHE, 2012b).This example demonstrates a need for greater campus conversation on what constitutes 'sustainability'.

One of the most effective avenues for holding such discussions exists in campus strategic planning processes. *PAE 2: Strategic Plan* recognizes institutions that have made a formal, substantive commitment to sustainability by including it in their strategic plans. Institutions earn the maximum of six points for including the economic, social and environmental dimensions of sustainability at a high level in the strategic plan, with two points awarded for each of the three dimensions. Analysis of data submitted for this credit as of 2013 showed that 79 per cent of institutions included all three dimensions in institutional strategic plans.

Interpreting sustainability holistically and focusing on the relationship between its social, economic and environmental dimensions is important for the development of a just and sustainable society. Cross-campus communication and inclusion of all dimensions in institutional planning processes are two effective ways to reach this level of understanding.

Identifying areas for improvement after submitting a STARS report

Once an institution has submitted its first STARS report, there is tremendous opportunity to engage the entire campus in reviewing the data, understanding gaps, conducting benchmark analysis, and devising strategies for improvement. By encouraging this level of dialogue, follow-up STARS reports are likely to show improvement in sustainability performance.

As of 2014, nearly 100 institutions had submitted a follow-up STARS report. STARS ratings jumped to the next rating level for 45 percent of these institutions, and average overall scores between first and second reports increased by eight points. Higher scores in follow-up reports may be the result of any of the following: (1) new initiatives and advancements in sustainability; (2) submission of a more complete follow-up report; and (3) greater campus-wide engagement after the initial rating.

When submitting a follow-up STARS report, obtaining campus-wide engagement is important for improving sustainability performance. While campus correspondents may not always be on board for collecting and submitting data for initial STARS reports, feedback from STARS liaisons in various formats has shown that, after an initial rating has been earned, campus correspondents tended to be more interested in collecting and submitting sustainability data for a second time. STARS provides an opportunity to engage the campus community in meaningful and continuous dialogue about campus sustainability at various stages and levels.

STARS as a tool for engaging students

A function of STARS that may not be particularly obvious involves its potential for engaging students in advancing campus sustainability. Since its release, students have been active supporters of STARS. Students were identified as key stakeholders early on, and their feedback continues to inform STARS technical development. Students experience STARS from multiple levels of engagement: (1) through active participation in the STARS reporting process; (2) through analysis of STARS data for student research; and (3) through enhanced sustainability awareness as a result of STARS rating recognition.

Engaging students in STARS offers a number of benefits to students and also for the institution. Perhaps the greatest benefit of engaging students in STARS is its potential for advancing sustainability on a broad scale. Students can actively participate in STARS in the data collection and reporting process, or they can use STARS data to conduct valuable research on campus sustainability. By earning a STARS rating, a buzz around campus sustainability can help identify new student engagement opportunities moving forward.

Post-submission survey data and webinar communication have shown that student experiences with STARS are overwhelmingly positive, and sustainability lessons learned can be applied throughout their lifetimes. In some cases, students began the experience with limited knowledge of what sustainability is about. Upon finalizing a report, students not only learned about the concept of sustainability, but had opportunities to apply systems thinking to address sustainability issues. As more students become engaged in STARS, the potential for advancing sustainability on a broad scale increases.

Student involvement in the STARS submission process

The STARS post-submission survey revealed that 'student involvement' in data collection for STARS was the third most popular data collection strategy (after 'working alone' and 'organizing a committee'). A 2010 AASHE webinar on engaging students in STARS found that involving them in the reporting process had numerous benefits for students and for institutions (AASHE, 2010). Students participating in the STARS reporting process develop a view of sustainability at the organizational level, while also drilling down into the details of energy, food, transportation, curriculum and other areas. Students also benefit through participation in a real-world work experience that enhances skills in data collection, analysis, reporting, interviewing, surveying, and presenting information. The experience provides students with opportunities for teamwork, collaboration, communication, leadership and self-directed, in-depth research.

The student engagement webinar also outlined ways in which institutions benefited from student involvement. Within higher education, students are the customers. As a result, communication from students to faculty and staff often resulted in increased cooperation and responses. At some institutions, student enthusiasm for sustainability led to greater cohesion within the STARS data collection team. The use of student time for sustainability reporting was also found to be less expensive than using faculty or staff to collect the data. Although students may sometimes be 'free' in real dollars, they are, however, likely to require more of another resource: time.

STARS institutions have used any one or a mix of three primary methods for involving students in STARS data collection:

- engaging student workers, interns and/or graduate assistants in data collection;
- STARS as a thesis project, independent study, capstone project, research practicum or related curriculum initiative;

- Student club involvement, or other co-curricular involvement opportunities.

Involving students in sustainability data analysis and research

Strong student research opportunities exist through student use of the STARS data displays and STARS API. Because STARS data can be easily accessed, students can conduct research on STARS even if their home institutions are not participating in the assessment.

Student research related to STARS can easily be identified within the Centre's Student Research database and Conference Presentations database. A keyword search of the word 'STARS' will deliver several student research projects and presentations, including some of those highlighted below:

- a doctoral dissertation from the University of Idaho that used STARS data to help interpret how sustainability literacy concepts are included in the curriculum of community college programmes (Rupple, 2012);
- students from the Northern Alberta Institute of Technology (NAIT) outlined new goals and initiatives identified through a student-led effort for submitting to STARS (Gareau and Avila, 2012);
- students from Duke University used STARS Investment subcategory data to help develop a model for encouraging sustainable endowment practices (Williams and Spangher, 2012).

For the Fall 2013 AASHE Annual Conference, several student presentations on STARS research were offered, including a student briefing on disciplinary trends in sustainability education based on STARS findings, and a doctoral student's dissertation on identifying correlations between performance and best practices using STARS.

Enhancing student awareness through STARS recognition

Students do not have to be deeply involved in the STARS submission process or in STARS data analysis to be impacted by STARS. The recognition of an earned rating provides a certain level of value for many students, often serving as a catalyst toward greater sustainability awareness. One of the documented outcomes of promoting a STARS rating is increased interest in sustainability among less-engaged students. This suggests that perhaps the best time to engage students in sustainability might be after an institution has earned a rating.

With competition as a strong driver of student engagement, students can provide unique perspectives on benchmarking and comparison. For example, students may be motivated to learn how their institution measures up in sustainability performance compared to its athletic conference rivals or regional competitors. This interest can promote campus sustainability in a general sense, and may also lead to new initiatives, such as a friendly inter-campus competition to reduce waste, energy and water use.

The current state of campus sustainability according to STARS

A wealth of information on trends and good practices is emerging as a result of the public nature of STARS data submissions. By communicating these findings broadly, advancements in campus sustainability can be achieved more quickly. Participation in STARS helps to answer the question, 'What are we learning about campus sustainability?'

More and more institutions are signing up for STARS and submitting valuable data related to campus sustainability. By 2014, over three hundred institutions of higher education had submitted a STARS report and earned a rating, and over 600 institutions were currently participating. STARS data are increasingly becoming the standard source for information related to all aspects of campus sustainability. Five years after its release, STARS can now be used to determine the current state of campus sustainability in the areas of curriculum, research, campus operations, administration and engagement. STARS data can identify important trends related to the campus sustainability transformation, but these data can also highlight areas for improvement within the reporting framework itself.

Outlined in this section of the chapter are key issues and constraints that have emerged as a result of STARS data analysis:

- Growth and expansion within the higher education sector is a significant obstacle for achieving sustainable development.
- A wide degree of interpretation in sustainability course identification may in part account for significant disparities in sustainability course offerings.
- Limited information is available on sustainability research among institutions where research is not considered during faculty tenure and promotion decisions.
- Separation of decision-making and lack of transparency make reporting on sustainable investment a challenge for many institutions.
- Institution type, research intensity and other institutional characteristics impact how institutions are doing in advancing sustainability.

Findings on campus infrastructure and sustainable development

An analysis conducted in 2013 and published in an AASHE blog highlighted STARS data and connections identified among building, climate and energy data points within the STARS Operations category (Urbanski, 2013).

In the Buildings category, a large disparity was found in certifications for newly constructed space and maintained space. More than a third of all new space earned a LEED rating for new design and construction, while only 1 per cent of maintained space earned a rating in LEED for Existing Buildings.[4] Conversely, 68 per cent of maintained space received no rating but operated in accordance with green standards, whereas 20 per cent of new building space received no rating but was designed in accordance with green standards.

As pointed out in a blog comment, one explanation to account for this disparity is that new buildings generally have a contract specification where documentation for LEED and submission for the rating are to be handled by contracts or architects. Existing building certification on the other hand must be undertaken internally or be contracted specifically. Other factors to consider in attempting to explain this disparity include the difficulty in attaining LEED ratings in both areas, value of each type of rating and financial constraints.

Perhaps the more important question is: 'How do these campus infrastructure trends affect the state of campus sustainability on a larger scale?' For most institutions, building operations tend to generate the greatest proportion of greenhouse gas emissions and energy consumption (ACUPCC, 2013). With this in mind, a review of climate and energy data should be carried out to provide additional clues on whether sustainable development is occurring.

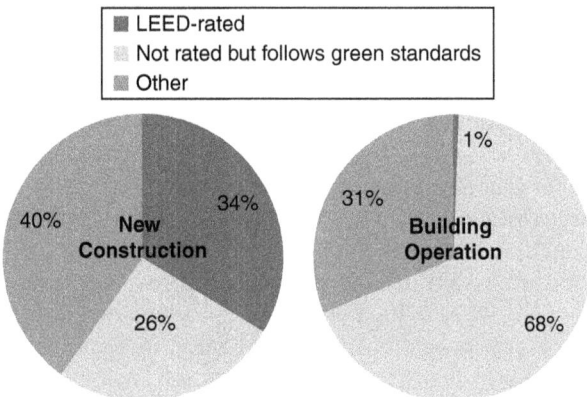

Figure 8.2 Analysis of STARS Version 1 data pertaining to maintained building space (*OP 1: Building Operations & Maintenance*) and new construction (*OP 2: New Building Design & Construction*)

Source: AASHE, 2013.

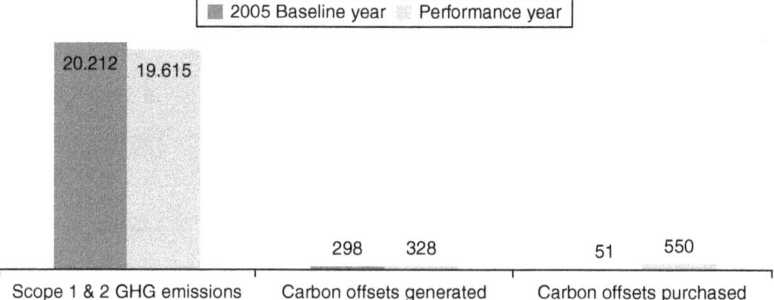

Figure 8.3 Total Emissions for all Rated Institutions under STARS Version 1.0 through 1.2, displayed in thousands metric tons carbon dioxide equivalent (CO_2e) in 2013
Source: AASHE, 2013.

Within the Climate subcategory, STARS data in 2013 revealed that between the baseline and performance year, average emissions per weighted campus user dropped from 6.5 to 5.5 metric tons of CO_2 equivalent. While a drop in emissions per weighted campus users may seem promising, it does not capture the true context of higher education's impact; it is after all the sum of all emissions that is found to be harmful to our environment and society.

STARS institutions reported that total scope 1 and 2 greenhouse gas emissions decreased slightly between the baseline and performance years, while carbon offsets generated on-site increased marginally. Purchase of carbon offsets grew ten-fold between 2005 and the performance year, though still covered only a fraction of total emissions. Despite a promising reduction in GHG emissions per weighted campus user and net increase in carbon offsets generated and purchased, this modest reduction in total emissions makes it clear that more needs to be done in the area of climate mitigation.

Analysis of changes in energy consumption and building space provides some interesting insights. In comparing energy consumption with building space, STARS institutions in 2013 reported a slight reduction in energy consumption over time. Average energy use dropped from 0.18 to 0.16 million British thermal units (MMBtu) per gross square foot (GSF) of building space between the baseline and performance years. However, a review of total energy consumption for all Rated institutions presents a less optimistic reality. Total energy consumption actually increased between 2005 and the performance year, despite a decrease in consumption per GSF. This slight increase can in part be attributed to a 14 percent increase in campus building space. The analysis reveals that a lot more needs to be done to achieve a sustainable rate of reduction in energy use.

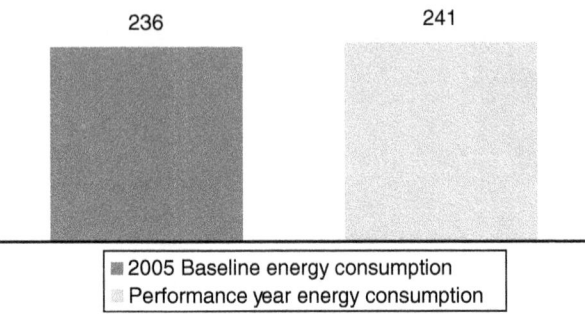

Figure 8.4 2013 analysis of change in energy consumption (in millions MMBtu) for all STARS Version 1 reports
Source: AASHE, 2013.

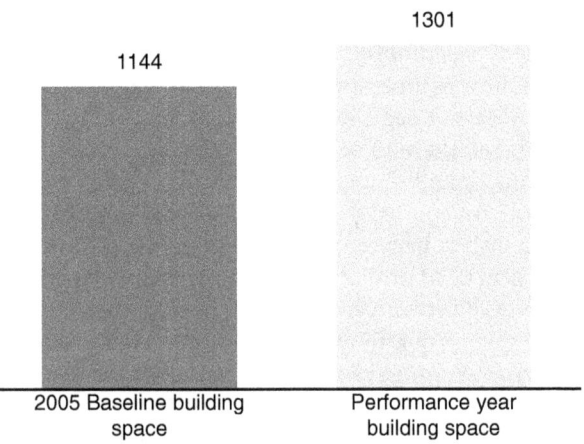

Figure 8.5 2013 analysis of change in building space (in millions gross square feet) for all STARS Version 1 reports.
Source: AASHE, 2013.

Despite a healthy pursuit of green certifications for new construction, growth and expansion within the higher education sector is a significant obstacle for achieving sustainable development.[5]

The challenge of measuring sustainability in the curriculum

The academic curriculum is arguably the most essential function for the mission of higher education institutions. Introducing students to sustainability throughout their college years prepares them for making responsible decisions and for potentially identifying sustainable solutions to today's most pressing social and environmental problems.

Through analysis of STARS data and feedback from STARS participants, we are learning that there are significant challenges for achieving standard sustainability metrics within the campus curriculum. According to STARS post-submission survey results, the Curriculum subcategory was identified as the most challenging area to complete within STARS. A wide degree of interpretation of what constitutes sustainability-focused and sustainability-related courses has been cited as a key challenge.

STARS Version 1 allows institutions to develop a definition of sustainability in the curriculum and allows flexibility in defining parameters for identifying courses that are sustainability-focused and sustainability-related. A 2013 AASHE blog outlined how allowing institutions to create their own definitions of sustainability in the curriculum resulted in a lack of comparability between institutions and greater likelihood of inflated course counts: 'Some institutions seemed to take a very inclusive approach to classifying "sustainability-focused" courses... while others took a more conservative approach, basically only including courses with "sustainability" in the title' (Dautremont-Smith, 2013). A great deal of scoring disparity exists for the two primary curriculum credits. In 2013, scores in *ER 6: Sustainability Focused Courses* ranged from 0.03 to 10, while scores for *ER 7: Sustainability Related Courses* ranged from 0.05 to 10. Much of this disparity is probably the result of differing interpretation rather than actual sustainability performance within the curriculum. Based on these and related findings, the STARS Steering Committee is working to incorporate additional guidance on defining sustainability in the curriculum for STARS 2.0.

Sustainability research opportunities for two-year institutions

Sustainability research is another area that requires careful observation and analysis, particularly as it relates to institution type. Within the current parameters of STARS, there is little incentive for non-research institutions, particularly associate colleges, to submit data related to sustainability research.

All credits within the Research subcategory apply to those institutions where research is considered during faculty tenure and promotion decisions. Based on these parameters, most two-year institutions may opt out of submitting data within the Research subcategory, with no adverse effects on scoring. By 2013, over 30 reports had been submitted by two-year institutions. Of these, only one college submitted information pertaining to sustainability research. With little data submitted, it is difficult to determine whether two-year institutions are opting out simply because research is not considered during faculty tenure and promotion decisions, or because sustainability research is not being conducted. Campus sustainability assessment needs to address how sustainability research differs based on institution type. Are there collaboration and partnership opportunities available for two-year institutions if standalone research initiatives are not pursued? Changes implemented in STARS Version 2.0 are expected to address applicability of sustainability research credits to encourage data collection from all institution types.

Box 8.3 Sustainability in curriculum, research and student engagement

Curriculum

- Two per cent of all courses were identified as 'sustainability-focused' by STARS institutions, while 7 per cent of courses were identified as 'sustainability-related'.
- Eighty-seven per cent of institutions offered a sustainability-focused undergraduate major or degree programme, while 83 per cent of institutions offered such a programme for graduate students.
- Twenty-one per cent of students graduated from programmes that had adopted at least one sustainability learning outcome.
- Twenty-one per cent of institutions conducted an assessment of the sustainability literacy of their students. Just over half of these institutions also conducted a follow-up assessment of the same cohort group.

Research

- Fifteen per cent of faculty members and 46 per cent of academic departments conducted research on sustainability topics.
- Seventy per cent of institutions had ongoing programmes to encourage student research in sustainability, while 59 per cent of institutions had programmes to encourage faculty sustainability research.
- Fifty-nine per cent of institutions gave positive recognition to interdisciplinary, transdisciplinary and multidisciplinary research during faculty promotion and tenure decisions.

Student engagement

- Forty per cent of institutions coordinate ongoing peer-to-peer sustainability outreach and education programmes for degree-seeking students.
- Ninety per cent of institutions held sustainability outreach campaigns that yielded measurable, positive results in advancing the institution's sustainability performance. Popular campaign initiatives often involved waste, energy and water use reduction.
- Seventy-six per cent of institutions included sustainability prominently in new student orientation activities and programming.
- Sixty-four per cent of institutions produced eight or more publications or outreach materials that foster sustainability learning and knowledge.

Source: STARS Version 1 Reports, AASHE, 2013

The challenge of addressing sustainable investment in higher education

The Planning, Administration and Engagement (PAE) category in STARS Version 1 focuses on institutional planning and administrative decision-making. This category addresses important social and economic dimensions of sustainability along with the more conventional environmental dimensions. While scores within PAE subcategories are typically among the highest within STARS, average scores in the Investment subcategory are significantly lower.

The Investment subcategory seeks to recognize institutions that make investment decisions that promote sustainability. Credits within the subcategory apply to institutions with endowments of US $1 million or larger. The STARS post-submission survey supports the finding that Investment is a particularly challenging area within STARS. Feedback from STARS participants showed that a particular challenge has been the separation of decision-making and lack of transparency in investment decisions. Within the US higher education system, boards or foundations that are responsible for investment decisions are typically not subject to federal, state or local freedom of information act requirements. As such, these entities do no typically share specific information pertaining to the investment portfolio of the university. As a result, many institutions found they were unable to collect the data needed to answer the Investment credits and consequently forfeited the entire category. Changes introduced in STARS 2.0 are expected to address transparency within the Investment subcategory to encourage institutions to submit investment data.

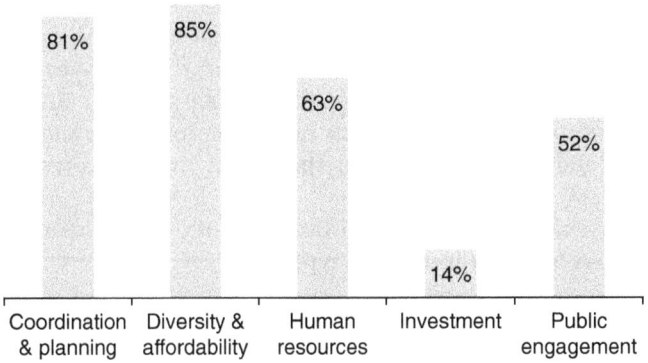

Figure 8.6 2013 Analysis of PAE Subcategory scores for all institutions submitting under STARS Version 1

Source: AASHE, 2013.

> Box 8.4 Sustainability in Planning, Administration & Engagement (PAE)
>
> - Sixty-five per cent of institutions had a sustainability plan and 67 per cent had a climate action plan.
> - Ninety-three per cent of institutions had a diversity and equity committee, office, or coordinator.
> - Seventy-seven per cent of institutions assessed attitudes about diversity and equity on campus and used the results to guide policy, programmes and initiatives.
> - Sixty-five per cent of institutions covered sustainability topics in new employee orientation and/or in outreach and guidance materials distributed to new employees.
> - Thirty-four per cent of institutions reported positive sustainability investments in one or more areas. However, sustainable investment holdings only represented 4 per cent of the total value of the STARS investment pool.
> - Sixteen per cent of institutions had a student-managed sustainable investment fund through which students could develop socially and/or environmentally responsible investment skills.
> - Ninety-seven per cent of institutions had developed partnerships with local communities to advance sustainability.
> - Sixty-five per cent of institutions had advocated for federal, state or local public policies that support campus sustainability or that otherwise advance sustainability.
>
> *Source*: STARS Version 1 Reports, AASHE, 2013

The impact of institution type on sustainability performance

Analysis of STARS data in 2012 has shown that institution type, size and research intensity impact how institutions are doing in advancing sustainability. When comparing average overall scores among all institutions based on these variables, it was found that four-year research institutions tended to earn higher scores than all other institutions. Four-year non-research institutions fell in the middle, while two-year associate colleges tended to earn the lowest overall scores (AASHE, 2012c).

The analysis in 2012 found statistically significant differences within subcategories based on institution type:

- Associate institutions scored lower than other institutions in Co-curricular Education, Curriculum, Transportation, Diversity & Affordability, and Public Engagement.
- Baccalaureate institutions scored higher than other institutions in Climate, Transportation, and Public Engagement.

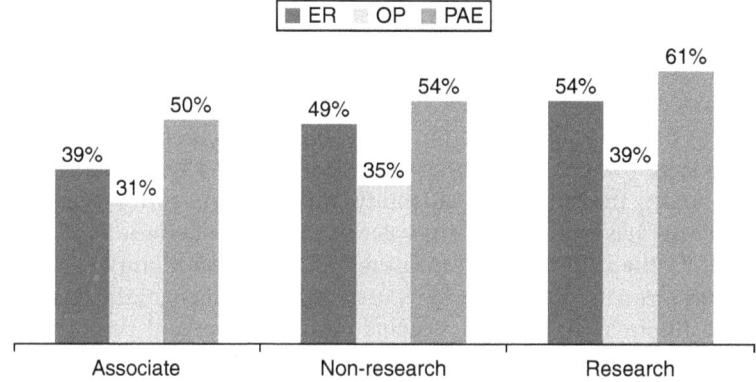

Figure 8.7 2013 analysis of average scores by category based on institution type for all institutions submitting under STARS Version 1
Source: AASHE, 2013.

- Master institutions scored lower than other institutions in Research, Climate, Grounds, Transportation, Water, Human Resources, Investment, and Public Engagement.
- Doctoral/research institutions scored higher than other institutions in Curriculum, Grounds, Transportation, Waste, Water, Diversity & Affordability, Human Resources, and Public Engagement.

Despite the overall trends, challenges and opportunities exist for every institution type. While research institutions struggle to reduce energy usage per square foot, they are often able to contribute significant financial resources for sustainability innovations. While small institutions may have fewer financial resources available for major investments in infrastructure, they are in a better position to integrate sustainability throughout the curriculum. Significant disparity in scores among large and medium-sized non-research institutions demonstrates that other, interrelated factors may impact sustainability performance. Ultimately, challenging areas can be identified for any institution, but these challenges can be overcome with planning and preparation. By using STARS as a tool for gap analysis and benchmarking, significant sustainability advances can be made each time a report is submitted.

Conclusion

At many campuses, the STARS rating has become a source of pride in demonstrating the campus's commitment to sustainability. Many campuses have created press releases and held celebrations around their STARS ratings.

Like any good assessment system, STARS has served to meet a wide variety of campus constituents' needs in determining how well they are doing in their sustainability efforts. The ability to be useful in numerous ways is largely due to the fact that the assessment has a high level of credibility among the campus sustainability community, in part, because this community has had a consistent voice in shaping it. Because STARS was shaped by the community, it measures what the community values, it provides positive recognition to its users, and it challenges the community to improve. Assessment systems need purpose, and the multiple purposes that STARS is meeting shows that measuring the quality of sustainability efforts is important to higher education institutions.

With significant impact on campus operations and social dimensions, STARS is helping to advance the campus sustainability movement through the five roles outlined in this chapter. While there is significant potential for STARS to serve a transformative role in the broader sustainability movement, quantitative, macro-level evidence to support this idea is currently limited. Future research on macro-level policy implications of STARS is needed to identify more clearly how STARS fits into the broad sustainability landscape. For example, how does STARS affect (versus reflect) what is happening in campus sustainability? Are there policy implications caused by STARS? What sustainability advancement potentials are being realized now that STARS is accessible to any higher education institution worldwide?[6]

As participation in STARS continues to grow, there is increasing potential for STARS to play a significant role in achieving a sustainability transformation across society as a whole. As more and more institutions participate in STARS, the collective impact for the broad sustainability movement will continue to grow. With new features released in 2013 that extend access to all higher education institutions on a global scale, this transformative effect may take shape in the very near future.

Notes

1. The STARS category topics were revised with the release of STARS 2.0. The Version One categories were Education & Research (ER), Operations (OP), Planning, Administration & Engagement (PAE), and Innovation (IN).
2. While STARS data displays do not currently include filters for climate zone, these types of data can be obtained from other sources. There are tremendous research opportunities for analysis of STARS data intersected with national or international climate statistics.

3. For scoring purposes, credits marked as 'Not Applicable' are excluded from the numerator and denominator when calculating final, category and subcategory scores.
4. While STARS 1.0–1.2 awards points primarily for LEED certification, STARS 2.0 is expected to extend to most international green building certification systems.
5. The Operations section of AASHE's Resource Centre provides specific recommendations on mitigation strategies related to energy conservation and efficiency, renewable energy technologies, maximizing space utilization, and designing and building only the greenest, most energy efficient new buildings.
6. See AASHE (2013, p. 11) for more information on enhanced access to STARS.

References

AASHE (2010) 'Engaging Students in STARS: A Student and Staff Perspective', http://www.aashe.org/files/documents/STARS/April%20Webinar%20PPT.pdf, date accessed 20 March 2013.

AASHE (2012a) 'Measuring Campus Sustainability around the World', AASHE Annual Conference, http://www.aashe.org/resources/conference/measuring-campus-sustainability-around-world, date accessed 12 May 2013.

AASHE (2012b) 'STARS Spring 2012 Quarterly Review: Framing Campus Sustainability', http://www.aashe.org/files/documents/STARS/sqr_spring_2012_final.pdf, date accessed 20 March 2013.

AASHE (2012c) 'STARS Quarterly Review Fall 2012: The Role of Institutional Diversity', http://www.aashe.org/files/documents/STARS/sqr_fall_2012-final.pdf, date accessed 20 March 2013.

AASHE (2013) 'STARS Annual Review 2013: A Look Back & A Look Ahead', http://www.aashe.org/files/documents/STARS/stars_2013_annual_review_final.pdf, date accessed 13 August 2013.

American College and University Presidents Climate Commitment (ACUPCC) (2013) 'Average Net Emissions by Source', http://rs.acupcc.org/stats/ghg-source-stats/, date accessed 17 May 2013.

Cortese A. (2003) 'The Critical Role of Higher Education in Creating a Sustainable Future'. *Planning for Higher Education*, 31 (3) 15–22. http://www.aashe.org/resources/pdf/Cortese_PHE.pdf, date accessed 19 May 2013.

Dautremont-Smith J. (2012) 'Benchmarking with STARS Data', Alfred State College, http://choose.esc.edu/suny-sustainability/wordpress/wp-content/uploads/2011/04/Julian-Dautremont-Smith_Alfred-State-College.pdf, date accessed 12 May 2013.

Dautremont-Smith J. (2013) 'How Many Dots to Connect? Defining Sustainability in the Curriculum Pt. 1', http://www.aashe.org/blog/how-many-dots-connect-defining-sustainability-curriculum-pt-1, date accessed 26 March 2013.

Gareau J., Avila M. (2012) 'SustaiNAITability', Northern Alberta Institute of Technology, http://www.aashe.org/resources/conference/sustainaitability, date accessed 19 May 2013.

Higher Education Associations Sustainability Consortium (HEASC) (2006) 'Call for a System for Assessing & Comparing Progress in Campus Sustainability', http://www.aashe.org/files/documents/STARS/HEASCcall.pdf, date accessed 12 May 2013.

Jaye S. (2012) 'Classification and Climate Zone Greenhouse Gas Inventory Benchmarking in Higher Education'. *International Journal of Facility Management*, 3 (1) http://www.ijfm.net/index.php/ijfm/article/viewArticle/46/62, date accessed 19 May 2013.

McIntosh M., Cacciola K., Clermont S., Keniry J. (2001) 'State of the Campus Environment: A National Report Card on Environmental Performance and Sustainability in Higher Education', National Wildlife Federation. p. 145, http://www.nwf.org/Global-Warming/Campus-Solutions/Resources/Reports/State-of-the-Campus-Environment-Report.aspx, date accessed 12 May 2013.

The Princeton Review (2013) 'College Hopes and Worries Survey Report', http://www.princetonreview.com/college-hopes-worries.aspx, date accessed 19 May 2013.

Rupple K. (2012) 'Education for Sustainability in Career and Technical Education: A Multiple Case Study of Innovative Community College Programs', University of Idaho, http://www.aashe.org/resources/student-research/education-sustainability-career-and-technical-education-multiple-case-stu, date accessed 19 May 2013.

Scott D. (2012) 'STARS Advanced Workshop: Strategies for Continuous Improvement', http://www.aashe.org/resources/conference/stars-advanced-workshop-strategies-continuous-improvement, date accessed 19 May 2012.

Shriberg M. (2002) 'Sustainability in US Higher Education: Organizational Factors Influencing Campus Environmental Performance and Leadership', University of Michigan, http://promiseofplace.org/research_attachments/Shriberg2002SustainabilityinHigherEdu.pdf, date accessed 12 May 2013.

Urbanski M. (2013) 'STARS Findings on Campus Infrastructure and Sustainable Development', http://www.aashe.org/blog/stars-findings-campus-infrastructure-and-sustainable-development, date accessed 20 March 2013.

Williams C., Spangher L. (2012) 'Endowment Transparency at Duke: a Qualitative Analysis of Effects on Environmental Sustainability and Assessment of Strategies', Duke University, http://www.aashe.org/resources/conference/endowment-transparency-duke-qualitative-analysis-effects-environmental-sustaina, date accessed 19 July 2014.

Part III
Quality Management and Facilitating Sustainability Competences and Capabilities

9
Sustainability and Values Assessment in Higher Education
Arthur Lyon Dahl

Education for sustainable development

The challenge of planetary sustainability is an urgent and complex issue for which appropriately trained experts and decision-makers are largely lacking. Higher education has been expanding rapidly in this field across a variety of disciplines including in the natural and social sciences, law, governance, international relations and even business, creating new challenges for assessment. In addition, the concept of sustainability includes an ethical dimension that has traditionally been seen to be difficult to assess. This short chapter reviews some of the complexities of sustainability as a subject for teaching and research in higher education, as well as some preliminary work on possible ways forward in the assessment of its values component.

Education for sustainable development at the university level faces a number of challenges, regardless of the discipline within which it is taught, which spill over into its assessment. The first is to define sustainability. There are multiple definitions, none of which is completely satisfactory. In fact, one of the advantages of the term, at least in a diplomatic context, is that everyone can read into it what they want, and that makes it easier to reach agreement. But what is a diplomatic advantage can be an assessment nightmare. How do you assess the understanding of a concept that is defined so loosely?

When the term sustainable development is used, it is also then necessary to have some common understanding also of what constitutes 'development'. Is it the 20th-century concept of helping the 'underdeveloped countries' to rise out of poverty and become modern economies? Does it only refer to economic development, as is commonly understood, or are there other dimensions to development? Does development equal growth? In the latter case, is development antithetical to sustainability, which can imply achieving a steady state, or even 'de-growth' for over-consuming societies? For many, the term sustainable development can constitute a contradiction in terms.

Despite these terminological and conceptual challenges, we have under way a UN Decade of Education for Sustainable Development (2004–2014), and many educational programmes at all levels. At the primary and secondary levels of education, the focus is more on education for environmental responsibility, sustainable consumption and responsible living. At the tertiary level, not only does teacher training need to include sustainable development education, but also all the disciplines from scientific and technical to high-level decision-making require education in the relevant understanding and skills necessary to achieve sustainable societies. Agreeing on the content to be taught and the capabilities and competences to be developed and then assessed is a challenge because of the diversity of approaches and requirements. Nevertheless, it is possible to identify some key components of any higher education programme in sustainability.

Sustainability

Sustainability requires a systems' perspective, providing an overview of how all the parts of the human and natural systems of the planet fit together and interact, and identifying the emergent properties of such complex systems, where the whole is more than the sum of the parts. The geographical spread is from local to global. Sustainability also requires a long-term view, since the range of time frames goes from the quarterly report in the corporation and the two- to five-year electoral cycle for politicians to decadal changes in the climate system and even processes operating on geological time scales. There may be great inertia in the large-scale systems that stretch out the time between cause and effect, or that link impacts across the planet. Sustainability challenges run across all of these scales in time and space.

Sustainability is also about dynamic processes, not a target to be reached but a balance to be maintained in space and over time (Dahl, 1996). It is traditionally defined as including economic, social and environmental dimensions, but many also add an additional dimension that can include institutional, cultural and/or ethical factors. Too often these dimensions (or pillars) have been treated separately and additively, with the environmental dimension dominant for scientists, and the economic dimension given the most weight by economists and politicians. In academia, as in government, it has been hard to break down the barriers between disciplines, and to get economists, for example, to collaborate with environmental scientists. Increasingly, however, it is acknowledged that sustainability in policy and action can only be obtained by integrating all the dimensions, and this is one of the challenges that has been set by the United Nations Conference on Sustainable Development (Rio+20) in Rio de Janeiro in June 2012 (United Nations, 2012).

Values and sustainability

Despite at least 40 years of international policy-making and action on sustainable development (even before the term was coined, when it was still called ecodevelopment), the world has continued on an increasingly unsustainable path. What are we missing? Most sustainable development education has emphasized scientific knowledge of the bio-geo-chemical systems of the planetary biosphere, and intellectual knowledge of the human economic and social systems and their challenges. But it is obvious in many contexts that intellectual understanding does not easily lead to changes in behaviour or lifestyle. Education also has to work at the level of motivation and emotional commitment, and this is more difficult both to implement and to assess. This may require challenging a person's basic assumptions and culturally determined preconceptions, for instance whether humans are naturally and inevitably aggressive and competitive, or whether education can bring out a higher potential for altruistic and cooperative behaviour (Karlberg, 2004; Novak and Highfield, 2011; Wilson, 2012).

Fundamental to this is an acknowledgement that sustainability is an ethical concept, aiming for justice and equity for all humanity at present and for future generations. The Brundtland Commission definition explicitly gives absolute priority to the elimination of poverty (WCED, 1987). Values, beliefs and ethics are a key driver for successful education for sustainability. Values make it possible to judge behaviour that benefits society. The individual operates on a spectrum from egotistical to altruistic, infantile to mature, base impulses to cooperative. In society this is expressed as power-hungry, seeking status and social dominance, versus conscientious, egalitarian, communitarian (Shetty, 2009). The latter values generally contribute to greater social good and higher integration. More than a decade ago the World Summit on Sustainable Development acknowledged the importance of ethics to sustainability in its programme of implementation (United Nations, 2002, para. 6). The question then is how to incorporate values into sustainable development education and assessment.

Values-based assessment tools

There has been extensive work on indicators of sustainability across its environmental, economic and social dimensions (Hak et al., 2007), but little to extend these assessment tools to cover the ethical foundations for achieving sustainability (Dahl, 2012).

To respond to this need, a European Union FP7-funded research project in 2009–2011 explored the development of indicators and assessment tools for civil society organization projects promoting values-based education for sustainable development (Podger et al., 2010; Burford et al., 2013; Podger et al., 2013). It brought together academic researchers from the University of Brighton (UK) and Charles University Environment Centre (Czech Republic) together with the Alliance of Religions and Conservation (ARC, UK), the Earth Charter Initiative (Sweden/Costa Rica), the European Bahá'í Business Forum (EBBF), and People's Theatre (Germany), with the collaboration of the International Federation of Red Cross/Red Crescent Societies (IFRC). These organizations are all involved in education at the level of values and ethics. The project looked for indicators that could measure the changes brought through these educational activities, and thus make values-based change more tangible. It was important that the project was led by the partner organizations, not the researchers, to avoid any imposition of a particular set of values. The organizations defined what values were important to them and what they wanted to measure, such as implementing values or spiritual principles. The researchers then helped to define assessment methodologies and indicators for the organizations to select and test in their projects, followed by a joint evaluation and sharing of experience. The indicators that were developed and tested were then shared more widely, leading to a final conference in December 2011 presenting the results of the project (ESDinds, 2011).

There were two major outcomes of the project. First, it helped the organizations to crystallize their own values. Some were explicit values in their charter or mission statement, others were implicit and only revealed through interviews with staff and participants. The researchers also reviewed different approaches to the definition and classification of values in the academic literature. Hundreds of terms for values were identified, but with little consistency, with wide variations in usage across different organizational contexts and disciplines, and difficulties in controlling for cultural bias. The project therefore adopted an empirical approach, letting each organization define its own values using terms that were meaningful in their own context. This process provided each organization with a common values vocabulary that it could use in its own activities, and ensured internal consistency within each assessment case study.

The project then selected a few broad values: unity in diversity, trust/trustworthiness, justice, empowerment, integrity, and respect for the community of life (the environment). These abstract concepts were not themselves directly measurable, but there was wide agreement on how they were expressed in various situations, such as in interactions among members of a group, attitudes or feelings expressed by individuals, or actions undertaken in implementation of a value. For example, within the general theme of trust and trustworthiness, 12 indicators were identified that reflected an atmosphere of trust (such as a safe space for sharing feelings or opinions, lack of gossip or back-biting, treating others with respect), a further 14 indicators covered the perception and presence of trust, 16 expressed actions that build and maintain trust, and 15 related to living by ethical principles as a foundation for trust.

The project narrowed down the lists to a common set of 166 indicators for attitudes or behaviours that reflected these values, as defined within the organizations, such as 'everyone has their place in the team', 'conflicts are resolved through dialogue', 'work is viewed as a form of service', or 'decision-making takes into account the social, economic and environmental needs of future generations' (ESDinds, 2012). These indicators had more general relevance than the vocabularies for values used in the different organizations and communities of practice. The same indicator might be useful for more than one value, or might measure what was expressed as different values in distinct organizational contexts. What was important was that the indicators as selected by the users were understood consistently within a particular project or organization.

A variety of measurement techniques were identified for use in the field. Some imagination was required to find techniques that were appropriate for indigenous school children in Mexico, former child soldiers in Sierra Leone, university students or business executives. Usually a combination of indicators and methods allowed for some cross-checking for consistency in results.

The indicators were tested in a wide variety of case studies as proof of concept: with the University of Guanajuato, Mexico (Earth Charter); Youth as Agents of Behavioural Change, Sierra Leone (Red Cross); Echeri Consultores, Mexico (Earth Charter); Lush Cosmetics, Italy (EBBF); People's Theatre, Germany; a Muslim women's group, London (ARC); and a financial services company, Luxembourg (EBBF) (ESDinds, 2012; Burford et al., 2013). The university project is particularly relevant to assessment in higher education, and will be used as an example here (based on ESDinds, 2010 and Burford et al., 2013).

Guanajuato University case study

The Environmental Institutional Programme of Guanajuato University (PIMAUG) is a cross-faculty initiative structured around six strategic areas:

1. Assisting students to develop a holistic vision of the environment;
2. Promoting sustainable resource use and waste management;
3. Diffusion of a culture of environmental awareness through a variety of media;
4. Interdisciplinary research;
5. Training in environmental issues through diplomas and master's programmes;
6. Social participation and inter-institutional partnership.

PIMAUG has a peer education programme in which Guanajuato University students train to run workshops inspired by the Earth Charter for the other students. It sponsors and coordinates a number of groups, such as the responsible consumer student group, the waste recycling student group, the habitat student group (dedicated to reforestation), and the group of staff coordinators of the environmental management system in each administrative and academic unit. Many of the students who participate in these programmes do so as part of the compulsory service element of their courses, for which they gain university credits, while others do so solely out of a desire to volunteer.

The university team was asked to review the full list of indicators (ESDinds, 2012) to determine which were seen as being relevant for the work they were doing. All 14 draft indicators for *Empowerment* and all 11 for *Trust* were validated as relevant by the PIMAUG group. Also validated as relevant were 6 of the 19 draft indicators for *Integrity*, 6 of the 8 draft indicators for *Justice*, 9 of the 12 draft indicators for *Unity in Diversity*, and 10 of the 79 draft indicators for the value of *Care and Respect for the Community of Life*. Only one indicator from the *Care and Respect for the Community of Life* value cluster and nine *Empowerment* indicators (three head indicators and six sub-indicators) were taken forward to the assessment stage in the pilot project.

The following indicators were selected for assessment in the project:

- People/partners become aware of how their existing knowledge, skills, networks, resources and traditions can contribute to the project/organization/team. Their contribution is encouraged, and people/

partners feel that their talents, ideas and skills have contributed to the outcomes of the project/organization/team.
- Workshop facilitators and participants are given autonomy and trust to fulfil responsibilities, at the same time receiving encouragement and support.
- Workshop participants are encouraged to express their opinion.
- The organization/team aims to provide all, especially children and youth, with educational opportunities that empower them to contribute actively to sustainable development.
- Individuals feel they are encouraged to reach their potential, and are provided with opportunities for personal growth.
- Individuals (a) develop programmes and deliver solutions on their own, and (b) have a sense of power that they can effect change.
- Work is viewed as a form of service (to the well-being and prosperity of all creation).
- People are given the opportunity to explore and reflect upon their own ideas and traditions, and then to develop their own vision and goals.
- People have identified their own responses to an issue, rather than just agreeing with the ideas of others.
- The project's activities/events produce an emotional connection to the community of life in participants.

A variety of assessment tools were used to collect data on the indicators, including:

- spatial and corporal surveys;
- semi-structured non-participative observation;
- focus group discussions;
- personal action plans;
- word elicitation – What/Why grid;
- key informant interviews.

The results fed back immediately into the work of the university team, with considerable impact.

As the university staff described it, the Earth Charter (ECI, 2000), which is a central focus of their work on sustainability values, is about transforming values into action. This is at the heart of the university's mission. The university already has good environmental measures, but there had been no way to assess the deeper dimension of the Earth Charter vision, and the degree to which its values were present and transformative. The selected indicators provided a way. They articulated deeply held aspirations and priorities which had not previously received systematic attention.

The PIMAUG team members found that the very act of reflecting on the indicators – even before associating them with specific assessment tools – allowed them to envisage new connections between their current activities, potential new areas of work that could be developed, and strategic decisions that they would like to take. The results of the assessment were also useful to PIMAUG in helping them to understand the efficacy of their workshops, identifying the factors involved in genuine empowerment, and providing insights into how motivation could be translated into effective action.

The culture of PIMAUG experienced a change through the assessment project. The Earth Charter workshop leaders reported a greater sense of effectiveness as a result of a clearer and more precise focus on values in their workshop delivery. The personal impact of the indicators affected how a manager dealt with conflict, and generated a much more participatory approach in her work with volunteers. The unit developed a greater unity of vision, and participants in the focus group discussions reported having reconnected or been re-inspired in their work. Integrating the indicators into regular evaluation increased group insight into their own application of values and led to understanding success in terms of values in a practical way.

The ESDinds project concluded with an international conference, 'Making the Invisible Visible: An Emerging Community of Practice in Indicators, Sustainability and Values' (16–18 December 2010, University of Brighton, UK), to present the results of the project to a wider audience, including educators, businesses, civil society organizations and social enterprises (ESDinds, 2011).

Implications for assessment in higher education

Sustainability education should aim to create an understanding of systems processes, both in the natural systems that determine planetary capacities and limits, and in the human systems that need to be managed to achieve sustainability. For the latter, ethical perspectives are an important part of human decision-making and choices of behaviour. The morality of sustainability (what is right or wrong) can have scientific dimensions (releasing carbon from fossil fuels into the atmosphere will acidify the oceans and destroy coral reefs) and humanistic/religious/spiritual dimensions (do unto others as you would have others do unto you). The science helps to provide a bridge between the abstract moral imperative and specific responses in terms of policy (for organizations) and behaviour (for individuals). The ethical dimension is one with high leverage. Behaviour driven by an internal ethical motivation will have a wider and more lasting impact than behaviour imposed by laws or regulations, and is much more cost-effective. These are strong reasons to include this dimension as a specific focus in educational programmes at all levels, hence the importance of assessment of educational results in terms of ethics and values.

Many universities are trying to 'green' their campuses and operations in an effort to show that they practise what they teach, and to attract motivated students. But that is only the material reflection of what should be a much deeper level of institutional and individual commitment to and transformation towards a sustainable future. Periodic surveys including values-based indicators could document evolving attitudes and behaviours among staff and students, and identify gaps requiring focused attention. As faculty across all disciplines come to understand the implications of the sustainability challenge for their fields, they will naturally begin to reflect this in their course design, and in interdisciplinary collaboration. The act of measurement itself, and the accompanying dialogue, creates an awareness of values and leads to their cultivation and application. This was apparent even in an ethically oriented university sustainability programme such as the case study cited above.

Student assessment in specific courses can also include dimensions beyond the simple assimilation of facts. Have the students acquired the tools of systems thinking in terms of processes and dynamic balance? Do they understand the complex interplay of natural and human systems, the risks of ignoring instabilities and tipping points, and the areas of leverage in human systems for maximum effect? Have they themselves formulated their own ethics of sustainability, which will be reflected throughout their lives? Has their education empowered and motivated them to use their newly developed potential to help solve some aspect of the sustainability challenge in their own field of endeavour? These are the kinds of questions that courses should address and that student assessment tools should aim to answer.

The experience acquired to date, including a decade of teaching sustainable development in higher education, suggests that including a values-based component raises the quality and impact of the educational process. The response, whether with graduate students or mid-career professionals in advanced studies, tends to be very positive. It helps to make what can be a large-scale, even depressing and de-motivating theme more immediately accessible and relevant at a human level, and suggests positive avenues forward.

Conclusions

This preliminary effort to assess the values dimension of education for sustainability has demonstrated that measuring behaviours or feelings linked to values in a scientifically valid way is possible. By agreeing to a common values interpretation within a programme or organization, the tracking measurements used can be given greater internal consistency and validity, generating indicators that can show the state of values or their change over time.

Indicators can make the values in a programme or educational activity more visible. When something can be measured, it becomes important. Values development can then be consciously encouraged or cultivated, making a programme or course more values-driven. Strong values are linked to more effective outcomes, so all human activity can benefit from stronger values.

The project discussed in this chapter used a variety of measurement methodologies that are sufficiently flexible to adapt to most situations (Burford et al., 2013). The approach can incorporate almost any values framework, as it is not prescriptive. The case studies showed that measuring desirable behaviours and values becomes positively reinforcing, and the partners are continuing to use the indicators in their work.

In higher education, it is clear that there is great potential to increase the values content of education for sustainable development, and to assess it in effective ways, both as the values are internalized in each student, and as they are externalized in programmes and activities. There is still considerable scope to develop practical assessment tools and procedures for use at larger scales than the small group or project activities used for the pilot projects.

With the explosion of sustainability challenges arising from climate change through food security to restoring social cohesion and transforming the economy, institutions of higher education need to be anticipating new careers and preparing students for them. In many cases, this will require overcoming disciplinary boundaries, and providing training that balances specialization and generalization. A values component will be essential in this, and new assessment approaches of the kind described here will help to put these new educational and research programmes on a sound footing.

References

Burford G., Velasco I., Janoušková S., Zahradnik M., Hak T., Podger D., Piggot G., Harder M. K. (2013) 'Field Trials of a Novel Toolkit for Evaluating "Intangible"

Values-related Dimensions of Projects'. *Evaluation and Program Planning*, 36 (1) 1–14.

Dahl, Arthur Lyon (1996) *The Eco Principle: Ecology and Economics in Symbiosis* (Oxford: George Ronald; London: Zed Books)

Dahl, Arthur Lyon (2012) 'Achievements and Gaps in Indicators for Sustainability'. *Ecological Indicators*, 17 (June 2012) 14–19.

Earth Charter Initiative (ECI) (2000) 'The Earth Charter', http://www.earthcharterinaction.org/content/pages/read-the-charter.html, date accessed 19 April 2014.

ESDinds (2010) 'Field Visit Summary: Guanajuato University, Mexico', http://www.brighton.ac.uk/sdecu/research/esdinds/resources/fieldvisitsummary_guanajuato.html, date accessed 19 April 2014.

ESDinds (2011) 'Making the Invisible Visible: An Emerging Community of Practice in Indicators, Sustainability and Values', Conference, 16–18 December 2010 at the University of Brighton, http://www.brighton.ac.uk/sdecu/research/esdinds/conference/index.html and http://iefworld.org/conf14.html, date accessed 19 April 2014.

ESDinds (2012) "Full Indicator List Developed from ESDinds Project', http://www.brighton.ac.uk/sdecu/research/esdinds/resources/indicatorlist.html, date accessed 19 April 2014.

Hak T., Moldan B., Dahl A. L. (eds) (2007) *Sustainability Indicators: A Scientific Assessment*, SCOPE Vol. 67 (Washington, DC: Island Press).

Karlberg M. (2004) *Beyond the Culture of Contest: From Adversarialism to Mutualism in an Age of Interdependence* (Oxford: George Ronald).

Nowak M. A., Highfield R. (2011) *Super Cooperators: Altruism, Evolution, and Why We Need Each Other to Succeed* (New York: Free Press).

Podger D., Piggot G., Zahradnik M., Janoušková S., Velasco I., Hak T., Dahl A. L., Jimenez A., Harder M. K. (2010) 'The Earth Charter and the ESDinds Initiative: Developing Indicators and Assessment Tools for Civil Society Organizations to Examine the Values Dimensions of Sustainability Projects'. *Journal of Education for Sustainable Development*, 4 (2) 297–305.

Podger D., Velasco I., Amézcua Luna C., Burford G., Harder M. K. (2013) 'Can Values be Measured? Significant Contributions from a Small Civil Society Organisation through Action Research'. *Action Research Journal*, 11 (1) 8–30.

Shetty P. (2009) 'Novels Help to Uphold Social Order'. *New Scientist*, 17 January, p. 10.

United Nations (2002) 'Report of the World Summit on Sustainable Development', Johannesburg, South Africa, 26 August–4 September 2002. A/CONF.199/20* (New York: United Nations). http://www.un.org/ga/search/view_doc.asp?symbol=A/CONF.199/20&Lang=E, date accessed 19 April 2014.

United Nations (2012) 'Report of the United Nations Conference on Sustainable Development', Rio de Janeiro, Brazil, 20–22 June 2012. A/CONF.216/16 (New York: United Nations). http://www.uncsd2012.org/content/documents/814UNCSD REPORT final revs.pdf, date accessed 19 April 2014.

Wilson E. O. (2012) *The Social Conquest of Earth* (New York: Liveright Publishing Corp).

World Commission on Environment and Development (WCED) (1987) *Our Common Future* (Oxford: Oxford University Press).

10
Educating Sustainability Change Agents by Design: Appraisals of the Transformative Role of Higher Education

Katja Brundiers, Emma Savage, Steven Mannell, Daniel J. Lang and Arnim Wiek

Introduction

While scholars observe positive trends in sustainability education, sustainability education as a field still finds itself mired between institutional inertia and strong drivers for transitions (Jones et al., 2010). As Van der Leeuw et al. (2012, p. 118) describe:

> Academic institutions remain so inertial because the professoriate remains in familiar and comfortable patterns. This is human nature, but denudes the academy of the energy and passion needed for change. Following form, the next generation of academics learns the habits, practices, and methods of their professors, replicating the status quo. A more bilateral relationship between faculty and students might produce different outcomes. If students played an equal role in the development of curricula, selection of course content, and initiation of applied projects, how different might the impact of the academy become?

The vision implicit in this description is of sustainability education defined by innovative, multilateral relationships among faculty, students and surrounding communities. This chapter presents work in progress at three educational sustainability programmes – one each in Canada, Germany and the United States of America – seeking to contribute to transformative change for sustainability by way of educating 'sustainability change agents' (Moore, 2005; Svanström et al., 2008).

The transformational potential of multilateral interactions can be realized though individual sustainability courses and through whole-degree sustainability programmes. Courses that adopt student-centred learning approaches expose students to real-world sustainability issues and bring them together with academic and professional experts.[1] Such Problem- and Project-Based Learning (PPBL) approaches train students in participatory methods and real-world sustainability problem-solving. Evaluation of such multilateral courses requires inclusion of all participants – students, experts and instructors – in evaluations of the collaborative process and its outputs throughout the course.

Scaling up from the course-level, whole-degree programme development and evaluation can also be tuned to multilateral interactions. One novel approach to the evaluation of such sustainability programmes is to implement an adaptive, collaborative evaluation uniting faculty, staff, students and members of the community. Critically reviewing the quality of such approaches and identifying means for improvement is essential because 'collaborative endeavors that target real-world problems, and...team participation in projects that accomplish assessment, assimilation, synthesis, implementation, and application' still struggle to be awarded equal prestige as fundamental research and traditional education (Crow, 2010, p. 488).

In this chapter we present participatory evaluation activities for collaborative sustainability courses and programmes. The evaluation activities offered aim to foster a culture of continuous development through quality systems that inform teaching and learning.

First we provide a general overview of the three sustainability programmes we draw upon. This will be followed by a section focusing on the course-level; in this section we present a framework to design and evaluate problem- and project-based courses for sustainability problem-solving followed by a section that applies this framework to a specific course for illustrative purposes. After the course-level we focus on the programme level; we present a framework for design and evaluation of a whole sustainability programme including a case on how this framework is applied in practice at Dalhousie University, one of the three programmes featured in the chapter. We conclude with a discussion of the achievements in implementing proposed design principles and comment on the distance to target and potential ways to close the gap.

Background information on sustainability programmes

The three programmes we discuss are the Environment, Society and Sustainability Major at the College of Sustainability at Dalhousie University, Canada; the Sustainability programme at the School of Sustainability at Arizona State University, USA; and the Leuphana University, Germany.[2] Although all three programmes offer undergraduate and graduate degree programmes, this chapter focuses on the undergraduate level.

In the past years, these programmes have admitted large cohorts of students, each averaging about 450 undergraduates in total and freshmen classes of about 140–200 students.[3] The faculty involved in these programmes is highly interdisciplinary; faculty hail from more than seven disciplines as well as experts from professional fields ('professors of practice'). All three schools consist of a core of 15–23 faculty members responsible for introductory and core courses, in addition to research duties. On average, some 12–40 affiliated faculty members contribute complementary course offerings. Student/professor ratios are around 25 students per professor. Between 30 and 50 funded graduate student teaching assistants aid with undergraduate classes. Support staff provide students with academic and career counselling (on average between five to eight advisors, student/advisor ratio: about 60 students per advisor).

To describe courses and programmes for sustainability we use the model of 'constructive alignment', because it provides a basic framework for designing and evaluating learning success (Biggs, 1999a). The 'constructive alignment model' combines a sequence of three steps: determining (1) *desired* learning outcomes, (2) learning and teaching *activities* deemed effective at facilitating attainment of desired learning outcomes, and (3) appropriate *assessment* schemes to evaluate student achievement of learning outcomes (Biggs, 1999a). The model emphasizes the integral interrelationships between design and evaluation of courses (Wiggins and McTighe, 1998).

Background on programme learning outcomes

The learning outcomes for each of the three sustainability programmes revolve around sustainability competences. Various scholars have worked on key competences in sustainability (cf. de Haan, 2006; Sterling and Thomas, 2006; Barth et al., 2007; Frisk and Larson, 2011). A literature review by Wiek et al. (2011) systematically compiled those contributions along five major competences:

- Systems thinking competence: 'the ability to collectively analyse complex systems across different domains (society, environment, economy, etc.) and across different scales (local to global), thereby considering cascading effects, inertia, feedback loops and other systemic features related to sustainability issues and sustainability problem-solving frameworks' (p. 207).
- Anticipatory competence: 'the ability to collectively analyse, evaluate, and craft rich "pictures" of the future related to sustainability issues and sustainability problem-solving frameworks' (pp. 207, 209).
- Normative competence: 'the ability to collectively map, specify, apply, reconcile, and negotiate sustainability values, principles, goals, and targets' (p. 209).
- Strategic competence: 'the ability to collectively design and implement interventions, transitions, and transformative governance strategies toward sustainability' (p. 210).
- Interpersonal competence: 'the ability to motivate, enable, and facilitate collaborative and participatory sustainability research and problem solving' (p. 211). As meaningful collaboration is a prerequisite for all other competences, the interpersonal competence is viewed as cross-cutting.

Wiek et al. (2011) highlight an overlap between sustainability competences and key components of sustainability research and problem-solving (see Figure 10.1). Sustainability research and problem-solving thus demands an *ability to design project teams that combine* all sustainability competences.

Background on learning and teaching activities

Following from the constructive alignment model introduced above, we provide a brief background on one activity tuned to the needs of sustainably education: problem- and project-based learning (PPBL). PPBL approaches afford rich learning experiences for students, and foster content knowledge and the development of a variety of metacognitive and practical skills (Blumenfeld et al., 1991; Hmelo-Silver, 2004; Savery, 2006). PPBL approaches are increasingly adopted in educational sustainability programmes in higher education (Stauffacher et al., 2006; Thomas 2009; Jones et al., 2010). While PPBL approaches are applied to some core courses of the programmes we draw upon, the majority of courses are offered in regular format.

Problem- and project-based approaches to education reflect a hybrid form, combining once distinct problem-based and project-based

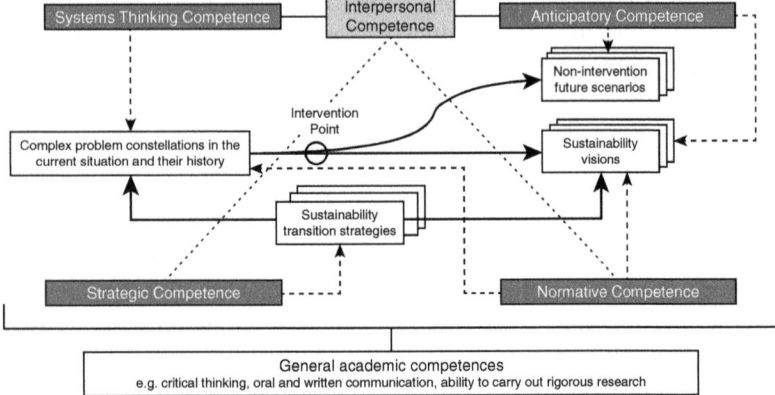

Figure 10.1 Illustration of how the five competences in sustainability (grey boxes) relate to components of sustainability research and problem-solving (white boxes)

Note: Dotted lines indicate how interpersonal competence cuts across competences; dashed arrows indicate relevance of individual competences for one or more components of sustainability research and problem-solving.

Source: Adapted from Wiek et al., 2011, p. 206.

pedagogies (Bereiter and Scardamalia, 2003; Donnelly and Fitzmaurice, 2005; Savery, 2006; Wiek et al., in press). At present the hybrid PPBL model promotes learning activities where students engage proactively in experiential learning processes by adopting a *self-directed* and *team-based approach* to work on *real-world problems and tasks*, simulating a *professional situation*. Students draw on and process *multiple information sources*, including reviewing secondary and collecting primary information. To support self-directed enquiry, instructors serve as facilitators of learning, rather than sole information providers. Instructors help students implement *formative assessments* to build capacity. Such formative assessments allow students to perform self- and peer evaluations and discuss with instructors ways to improve learning and collaborative abilities.

Background on PPBL course evaluation

Concluding the constructive alignment model introduced above, we provide here a brief discussion of ways to evaluate the PPBL activity described above. Two special features of PPBL courses are that they combine formative and summative assessments (Hmelo-Silver, 2004), and that these assessments involve the instructors, the students and peers, as well as partnering stakeholders (Brundiers and Wiek, 2013).[4] The goal of summative assessments is to ascertain the practical relevance of project outcomes; the goal of formative assessments is to ascertain students' learning achievements and strategies. Combined, the formative and summative assessments help determine ways to improve project outcomes as well as learning approaches (Nicol and Macfarlane-Dick, 2006).

Both forms of evaluation in PPBL courses draw on students' self-reflective journal entries and peer evaluations as grade-contributing performance assessment of team members. Additional evaluation formats that test real-time performance of students, for example their problem-solving ability, may also be incorporated (Biggs, 1999b). Rubrics are often especially helpful for evaluations, as they help students with self-assessment and instructors with formal evaluation of students' success without providing too much guiding structure (Knight, 2006; Ash and Clayton, 2004).

Course-level: design, analysis and evaluation framework for PPBL courses

In this section we present a PPBL course design, analysis and evaluation framework for sustainability education. The content of this section is abridged and adapted with permission from the article 'Do We Teach What We Preach? An International Comparison of Problem- and Project-based Learning Courses in Sustainability' (Brundiers and Wiek, 2013). The original article was published as part of a special issue, 'Sustainability in Education: A Critical Reappraisal of Practice and Purpose', in the journal *Sustainability*. For the purpose of this chapter, we draw on the analytical-evaluative framework for project- and problem-based courses in sustainability programmes.

Sustainability problem-solving as educational practice

The transformational branch of sustainability science goes beyond analysing and describing sustainability problems by developing the knowledge and techniques necessary to create sustainability solution options to help resolve sustainability problems (Spangenberg, 2011; Wiek et al., 2012). This perspective attracts many students who apply to sustainability programmes. The PPBL approach in sustainability programmes promises students opportunities to develop important lifelong learning skills in general and sustainability problem-solving competences in particular (Stauffacher et al., 2006; Brundiers et al., 2010). Despite this major draw for students, few studies have systematically examined how PPBL formats help build sustainability competences (e.g. Ferrer-Balas et al., 2008; Segalas et al., 2009). Examining the extent to which programme-level and course-level learning outcomes deliver on their promises is vital to meeting the objective of sustainability science and education vis-à-vis recruiting and training talent to contribute to sustainability problem-solving. As more in-depth evaluations are developed, early appraisals of how sustainability programmes meet objectives can help with programme and course improvement.

Brundiers and Wiek (2013) offer one analytical-evaluative framework for such attempting appraisals. The authors developed the framework on the basis of a literature review integrating principles, concepts and methods from sustainability science, education and transdisciplinary collaboration. The authors applied the framework by appraising six case studies.[5]

The framework builds on the research phases for solution-oriented sustainability research. The four main phases of this research include *orienting, framing* and *doing the research* as well as *implementing* the research results, referring to implementing the solution option (Wiek and Lang 2011). Applied to an educational course setting, those four phases are framed by a pre-course *preparation phase* and a post-course phase (*post-course extension*). Figure 10.2 illustrates the resulting five phases: preparation, orienting, framing and doing the research, as well as post-course extension and implementing the solution option phases.

The framework provides for each phase an analytical design question as well as an appraisal question. These questions were derived from the literature in the pertinent fields of sustainability science, education and participatory research (see Table 10.1). In the remainder of the section, we review each phase of PPBL course design.

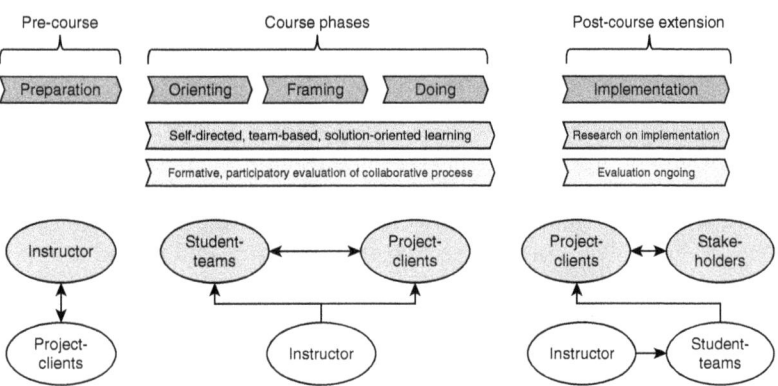

Figure 10.2 Five phases for problem- and project-based, solution-oriented sustainability research courses and project-participant constellation per phase, with phase-lead in grey

Source: Adapted from Brundiers and Wiek, 2013.

Table 10.1 Analytical-evaluative framework adapted from Brundiers and Wiek (2013)

Analytical design questions	Appraisal questions
Preparation and Orienting Phase	
1. Who is involved in the problem definition and result formulation?	1. Do project partners equally agree on the relevance of the problem and the research objectives?
2. What is the problem?	2. Is the problem defined as a *sustainability* problem?
3. What is the goal of the project?	3. Is the goal to develop actionable solution options?
4. What are the learning objectives of the PPBL course?	4. Are the learning objectives linked to key competences in sustainability? Are learning objectives individualized?
5. How are teams composed? How is teambuilding organized?	5. Do the teams account for expertise and interests? Are teambuilding techniques (e.g. code of cooperation) used?
6. How is PPBL as a learning- and teaching environment introduced?	6. Is there an explicit introductory PPBL tutorial with provision of resources, tools, and techniques?
Framing Phase	
1. What research methods are selected and combined?	1. Is a solution-oriented methodological framework adopted or developed? Is sufficient time allocated to each module?
2. Are participatory settings determined?	2. Do the participatory settings reflect the project objectives, as well as expertise and interest of the participants?
Doing Phase	
1. How is the research conducted?	1. Is research conducted according to the methodological framework created? Are all methods applied according to quality standards?
2. What are the ultimate research results?	2. Is a solution option developed based on the modular results? Does an extended peerreview inform credibility and saliency of the results? Are insights generalized beyond the specific case?
3. Are process evaluations performed?	3. Are formative evaluations conducted? Do students reflect on their experience and the quality of process and products? Do participants feel that agreed upon expectations were met? Are evaluation results implemented?
4. How is the acquisition of sustainability competences ensured?	4. Are students provided adequate support in developing sustainability competences?
Implementation and Post-course Extension Phase	
1. What happens after the main research is completed?	1. Is implementation of the research results moving forward? Is research on implementation lined up?
2. Who is involved in the implementation phase?	2. Are students involved as part of their overall PPBL experience?

Preparation phase: identifying the problem and project clients

For course preparation, three key planning areas emerge from the literature on education for sustainability (Stauffacher et al., 2006; Rowe, 2007; Brundiers and Wiek, 2011): project topic, project team, and instructor and student resources.

The project topic planning area deals with questions like: what is the sustainability problem or solution option that students address? Who identifies this as a sustainability problem or solution option respectively? What are desired outputs that clients and stakeholders envision with respect to addressing the issue? What learning outcomes in terms of sustainability competences does this project afford? Ideally students work on real-world problems, which are brought to sustainability programmes in higher education by community members, including representatives of businesses, public agencies or civil society organizations (Rowe, 2007). For students, the grounding of research projects in real-world context enriches their learning through the 'pedagogy of place' (Gruenewald, 2003). A first step for course instructors and project clients is to jointly determine the sustainability dimensions of the real-world issue and ensure that sustainability learning outcomes for students are compatible with desired project outcomes.

The project team planning area deals with questions like: Who works on the project? What learning approaches are used? The complexity of real-world sustainability issues and the limited amount of time to work on the project necessitate teamwork. Generating credible knowledge for sustainable solution options requires that project partners and potential additional stakeholders are able to participate in ways that they perceive as meaningful, appropriate and legitimate (Cash et al., 2003). Therefore, it is important for instructors and project partners to discuss possible participatory research settings and methods, key roles and responsibilities for all parties involved, and their implications for the course and research results (see Figure 10.2).

The resources needed for instructor and students planning area deals with questions like: What information, contacts or trainings would instructors and students need in order to succeed with the course and project? PPBL approaches are self-directed by students, requiring profound shifts in mindsets of students and instructors. Students need to shift from being a receiving learner to a self-directed learner, which in turn asks instructors to shift from providing information upfront to becoming a facilitator of learning and a resource guide (Donnelly and Fitzmaurice, 2005). These behavioural changes require conscious preparation and continuous practice (Savery, 2006). The syllabus can support this shift for students and instructors by pre-structuring the project and describing intermediate deliverables, deadlines and expectations. If the project and process are too unstructured, students can feel overwhelmed and frustrated, which has the potential to undermine the learning environment (Roessingh and Chambers, 2011).

Course phases: orienting, framing, and doing the research

Once preparation is complete, the *orienting phase* entails students learning about the research projects and the nature of PPBL courses. When orienting student teams to research projects, discussions around research objectives and questions need to delve into (1) the structure of the problem as a sustainability problem, (2) the anticipated project results (i.e. the sustainable solution options accounting for sustainability and the needs of project clients; Gibson, 2006; Wiek et al., 2012); and (3) the sustainability competences being developed (Wiek et al., 2011). When orienting students to PPBL pedagogy, class discussions need to address key aspects of the philosophy and process of PPBL, and student teams need to engage in a series of activities to become familiar with the resources, tools and techniques that enable success in PPBL environments (Moust et al., 2005).

Once oriented, students are prepared to frame their specific research project. The goal of the *framing phase* is for students, in consultation with their project clients, to identify a methodology for producing actionable knowledge to address the research problem. Actionable knowledge is a combination of three knowledge types needed to create clear instructions for how to carry out sustainability-oriented interventions for problem-solving: (1) *descriptive-analytical* knowledge about the past, current and future states of the problem from a systems perspective; (2) *normative* knowledge, assessing the extent of the problem and envisioning sustainable future states using relevant sets of values, norms and thresholds; and (3) *instructional* knowledge that develops intervention and transition strategies, and designs and tests policies, or develops organizational change programmes to resolve or mitigate the identified problem (Wiek et al., 2012). Each of these types of knowledge are generated with different families of methods and correspond to the sustainability competences (see Figure 10.1).

In addition, the framing phase involves laying a foundation for formative and participatory evaluation in the PPBL course (Blackstock et al., 2007; Brundiers and Wiek, 2011). Such evaluations are multifaceted and include self-evaluation, as well as evaluation of the performance of peers, instructors and project partners in the collaborative process. Course instructors need to support teams in this process by setting aside time for regular reflections and requesting individual preparations for those sessions, for example through structured journals (Hmelo-Silver, 2004).

Finally, in the *doing phase* students conduct the research they have collaboratively framed. Extended peer review processes with project clients and other experts are recommended mechanisms for short time frames to test the promise of developed sustainability solution options. A 'closure' meeting should outline how the same or a new cohort of students can continue the research through, for example, evaluative research on the implementation phase. These processes provide students with opportunities to practise professionalism and develop specific sustainability competences (Meehan and Thomas, 2008).

Post-course phases: implementation and post-course extension

Implementation is the process of applying the research results and realizing the solution option, which is the responsibility of project partners and other stakeholders. Research suggests that involvement of researchers in the implementation phase supports continuation of the process (Potvin et al., 2003; Bammer, 2005; Fraser and Galinsky, 2010). From an educational perspective, scholars argue that such *evaluative research on the implementation* of the solution option should be a goal of the PPBL course from the outset (Rowe, 2007). Hence, the goal of the post-course extension is to line up opportunities for students to perform evaluative research on implementation (e.g. through internships or independent research).

Course-level: applying the framework for PPBL courses

To illustrate the usage of the analytical-evaluative framework for designing and evaluating PPBL courses that help building key sustainability competences, we detail an exemplary case referenced in the paper by Brundiers and Wiek (2013).

Course description

We conducted a course appraisal of a one-semester long PPBL course: 'Don't Landfill it! Transforming the University's Food Waste System', offered at the School of Sustainability. The course was co-taught by an adjunct faculty member and the community–university liaison of the school. The project clients were three operation managers of the university: the directors of the University's Sustainable Practices Network; Ground Works, and Solid Waste. Sixteen upper division undergraduate and six graduate students participated in the course. The questions provided in Table 10.1 are used to describe the course and conduct the appraisal.

Preparation phase: identifying the problem and project clients
The need for a sustainable food disposal practice became the inspiration for the 'Don't Landfill it' course. The university's research need was translated into a course through a meeting with the director of the university's

office of sustainability practices and the school's community–university liaison. They clarified their expectations and the responsibilities of all parties involved in a Memorandum of Understanding (MoU). The memorandum allowed for a sparsely predetermined project frame and curriculum, granting students some autonomy to develop specific projects. The memorandum specified three components of the results of the course: (1) a vision of a sustainable food waste system; (2) a mapping of the legal, technical and social barriers and opportunities to implementing such a system, and (3) a strategy for implementation of the sustainability vision.

The MoU also defined the project and educational goals. The *project goals* were to develop a place-based, sustainable vision of a food waste system, perform an assessment of the current state vis-à-vis this vision, and draft a strategy to implement the vision. The *educational goals* were to expose students to all five key competences and offer opportunities for practising related skills. Developing competences entailed preparing students to be able to (1) explain the sustainability competences to others (e.g. potential employers), (2) evaluate personal strengths and weaknesses related to each competency, and (3) identify activities and documents of the course that demonstrate attainment of sustainability competences they wanted to focus on in this course.

The MoU held that clients and course instructors acted as co-leaders of the project in order to integrate the need for a practical, economically viable and a low-risk sustainability solution option with the goal of generating rich educational experiences for students. Project clients' responsibilities comprised sharing their experience and everyday work environment with students, critically engaging with students' works-in-progress, and acting as conduits to some 20 additional stakeholders in their respective fields. Additionally, project clients met regularly with instructors to review progress and discuss next steps.

To foster team-based learning and peer-mentoring, students collaborated across interconnected teams (Figure 10.3). Undergraduate students first self-organized into three thematic groupings of five students. Two graduate students led each thematic group. Second, three undergraduate students – one from each theme – joined a graduate-student-led synthesis team. Synthesis teams researched the food waste systems of six universities known for pioneering work in sustainable food waste management. A key benefit of this organizational structure was having team members from each thematic group inform and be informed by the case study analyses. Finally, students were responsible for individual assignments related to building capacity for teamwork.

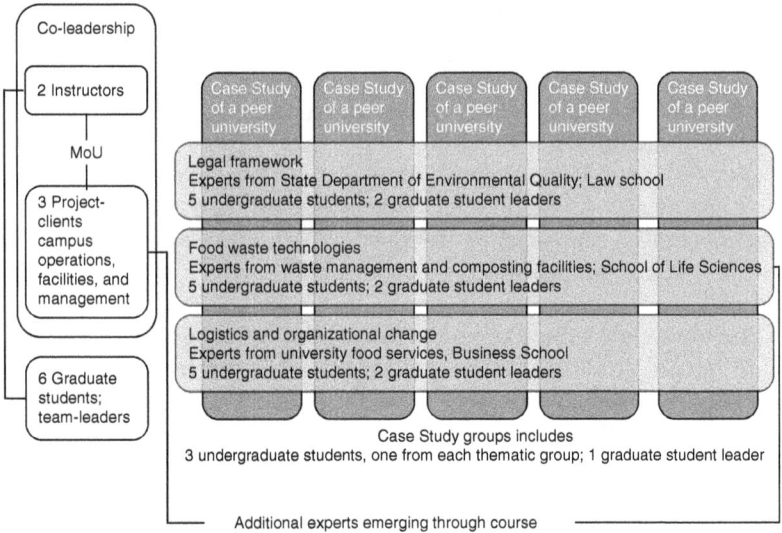

Figure 10.3 Organization for team-based courses to support collaboration with project clients and peer-learning

Note: Legal framework, social change management and composting technologies represent thematic groups; case studies one through five represent synthesis across teams.

Course phases: orienting, framing, and doing the research

Orienting phase: In the first weeks of the course, students formed teams. To equip the teams, instructors introduced the basic steps of PPBL, facilitated a class discussion on key competences in sustainability, and asked the student teams to develop their 'tools for teams', for example a 'contract of collaboration' defining the rules for collaboration. In addition to team formation, students toured the campus to explore opportunities for sustainable food waste management. Students were challenged with framing food waste management as a sustainability problem using sustainability criteria compiled by Wiek et al. (2012), identifying potential intervention points and researching existing solutions strategies. Based on this research, along with project clients' perspectives, work experiences, available data and desired objectives, student teams developed final research objectives and questions that were approved by project clients.

Framing phase: Once oriented, student teams were tasked with drafting work-plans to describe project objectives and approaches, listing responsibilities and action items for each team member. Main project activities were delimited by course instructors and included: a current state analysis, including a waste audit and a sustainability assessment; a visioning process to develop a sustainable food-waste management system; and a strategy development process involving focus groups and campus staff. Team workplans had to demonstrate compliance with ethical research requirements and coordination among student teams about contacting stakeholders.

Doing phase: Students began conducting research in the third week of the semester. Research activities included conducting a day-long waste audit to collect, separate and measure food waste; developing and administering surveys of peer-universities to learn about sustainable food waste management strategies; visiting composting and waste management facilities; and interviewing experts and conducting focus groups related to team-specific thematic areas. Project products included a final event for project clients, stakeholders and the interested public; project summaries for public audiences; and a comprehensive report for project partners. Research results included: (1) strategies to create buy-in and build capacity for the proposed food waste management programme among kitchen and cleaning staff and among consumers (students, faculty and staff); (2) a proposed green-bin collection method; and (3) a combination of composting methods, including small-scale educational demonstration projects and large-scale vessels operable either by the university or an external partner.

Throughout the semester, instructors and graduate students met in weekly supervision meetings. Midway and at the end of the semester, all students evaluated team processes and the course using instructor-provided questionnaires. Questionnaires explored general satisfaction with the process of the course and asked students to document learning gains and explain perceived challenges related to teamwork, self-directed learning, and sustainability competences.

Post-course phases: implementation and post-course extension

The project ended with the final event and a debriefing among instructors and project clients. While no student activities were lined up to continue the research after the course, two project partners continued with the work developed with the student teams. The solid waste manager pursued the idea of the green-bin and the university sustainability practices office continued with the exploration of a sustainable food waste system. In addition, an honours undergraduate student and a graduate student have since further advanced the research initiated by the course.

Course appraisal using the analytical-evaluative framework

With the exemplary course described, we now briefly appraise how well the course design matched implementation. Thereby, we use the appraisal questions listed in Table 10.1. We focus on those aspects of the analytical-evaluative framework that address key challenges for PPBL in sustainability courses, as identified by Brundiers and Wiek (2013). These challenges relate to project framing, team-based learning, and participatory approaches:

1. *Project framing*: framing projects as sustainability projects in order to allow students to practise sustainability competences related to knowledge, skills and attitudes;
2. *Team-based learning*: acknowledging that self-directed and team-based learning (as part of the interpersonal competence) are new to most students, instructors and project partners and therefore require introductory activities to practise key skills;
3. *Participatory approaches*: adopting a differentiated approach to participatory methods with project clients and other experts involved in the research process.

Appraisal of project framing

The project offered students opportunities to practise key sustainability competences. Students framed projects as sustainability projects using formal sustainability criteria (e.g. Gibson, 2006; Wiek et al., 2012). All project clients committed to the project goal of developing an actionable solution option, which resulted in a proposed strategy for a sustainable food waste system. The continued post-course engagement of university staff suggests that the solution option proposed by the student team was *actionable*. In the framing phase, students specified a partly predetermined research methodology, which combined a *solution-oriented set of methods* related to the knowledge types necessary for developing solution options, including descriptive-analytical, normative and instructional knowledge. Students dedicated about *equal amounts of time to each research module*. This compact approach caused problems for undergraduate students lacking prior exposure to basic research skills and solution-oriented research methodologies.

The *learning objectives* for this course were linked to key competences in sustainability. In the summative evaluation, the majority of students indicated appreciation for the opportunity to work on a real-world problem. Further, students appreciated working along all of the components of a solution-oriented project because it afforded them a sense of 'how it all fits together' and how each of the five sustainability competences translate to knowledge, skills and attitudes. Nevertheless, meeting this learning objective was hampered by instructors failing to ensure that student teams contained individuals with strengths across the sustainability competences (students self-organized into teams). Furthermore, the instructors missed the opportunity to combine the initial self-assessment with a post-course formal assessment of sustainability competences.

Appraisal of team-based learning

While this PPBL course offered opportunities to practise teamwork and self-directed learning, instructors could have offered more practice opportunities. The course instructors introduced PPBL principles and methods and asked the graduate student team-leaders to deepen those introductions with their undergraduate teams. The majority of students reported that they found it helpful to be required to use a set of teambuilding and team-working tools. For instance, the results of the mid-term peer evaluations and mid-term course evaluations enabled a conversation within teams and between students and course instructors about the effectiveness of the course in supporting teamwork and self-directed research. Graduate students reported the weekly meetings with peers and course instructors as helpful for supporting their undergraduate peers in productive, fair and reliable teamwork and self-directed learning. Instructors used the information from the weekly meetings and mid-term evaluations to adjust the course and act as mediators for teams in times of tensions. Despite these positive reports around work-oriented activities, students reported that introductory teambuilding activities would have benefited from additional *social* activities, such as ice-breaker games, a potluck, and more fieldtrips with the entire class. If such activities are adopted in the future to ensure teambuilding across the class, these activities should also include project clients.

When asked about the one thing that students would change to make the course better, almost half of the students expressed a need for clearer expectations around timelines for deliverables. While we empathize with these student impressions, we note that some level of uncertainty goes hand-in-hand with self-directed pedagogies; research needs to emerge through research activities and important decisions need to be made often when lacking key aspects of knowledge. The gap between student expectations for certainty and the realities of uncertainty necessitates that instructors may enhance structure where possible, but most importantly should reinforce the normalcy of this tension and discuss constructive coping strategies.

Appraisal of participatory approaches

The PPBL course missed the opportunity to go beyond extractive research methods in its stakeholder engagements. Although project clients visited the class three times and engaged with students during those meetings, communication remained on the level of information exchange, and did not advance to collaborative co-creation of knowledge. Only in the 'doing-the-research phase' did students begin to specify stakeholder engagement methods. Given the time needed to plan collaborative workshops, the result of this late-phase initiation were stakeholder engagements involving one-off or one-way data collection methods (e.g. surveys, focus groups, presentations). The originally hoped for two-way collaboration did not occur, in part owing to a lack of knowledge among students and in part because of timing constraints. Instructors need to better anticipate the varying levels of student knowledge of participatory methods and reinforce the need for early engagement to account for time constraints.

In feedback sessions between project clients and students at mid-term and final presentations, discussions revolved around credibility and saliency of the project, without touching on the *process of engagement* (legitimacy criterion). By contrast, project clients and instructors met periodically to discuss process and share constructive feedback on the projects. These various interactions revealed two surprising insights: (1) Classes offer project clients a safe environment for collaboration *among themselves*; while project clients' visits to the class were framed as opportunities for communicating information *to students*, the project clients experienced the value of using the classroom as a venue to communicate indirectly with their counterparts. (2) Students perceived project clients as reticent and critical. What students took for reticence, project clients explained as a strategy to maintain their open-mindedness for *learning from students* without imposing jaded perspectives. Instructors can do more to facilitate mid-term and final reflections on process components of engagement, and explore with project clients how to balance students' requests for open communication without colouring students' perspectives.

The final public event showcased students' work and raised interest among campus staff, but did not catalyse *immediate* steps to implement the proposed actions, which caused dismay among students. In part this is an opportunity for instructors to learn to better manage expectation about change in university settings. However, this also reflects a need for students and instructors to prepare early on for ways to continue projects post-course. These conversations can teach students about the realities of reform in bureaucracies, and how to plan accordingly.

In sum, the semester-long workshop course 'Don't Landfill it! Transforming the University's Food Waste System' responded to a societal need, addressed a sustainability issue, was framed as a solution-oriented project, and resulted in a sustainable solution option (proposal for sustainable food waste management). Although the course failed to plan for the continuation of the project, campus staff and students have since carried the project forward. A majority of students appreciated team-based and self-directed learning as an opportunity to develop interpersonal skills; more than 40 per cent of students thought that *learning about* a solution-oriented methodology informed their understanding of sustainability competences. Instructors and project clients entertained open communications, which enabled mutual learning on how to improve collaboration in the future and mentor students better in enhancing their capacities and skills.

Having discussed implementation of course-level design and evaluation, and conducted an appraisal of an exemplary course, we turn now to a discussion of programme-level evaluation.

Programme-level: developmental evaluation framework

New emergent programmes in education for sustainable development are taking shape in numerous formats and capacities. There is a growing body of academic literature that offers approaches to quality assurance and improvement in programme design and teaching and learning environments for sustainability in higher education. Wiek et al. (2011) compile the key sustainability competences of sustainability change agents. Yet, a gap exists between how these competences translate into measureable programme-level learning outcomes. As these new sustainability programmes emerge and grow in number, maintaining academic standards and quality assurance through rigorous evaluation is becoming increasingly important (Shriberg, 2002; Bornman, 2004). We propose that developmental evaluation is best suited to guide sustainability programme development by using evaluative data to support creativity, social experimentation and adaptation throughout the programme level.

Principles of a development evaluation

Developmental evaluation (DE) describes a collaboration between social innovators and evaluators, a co-created and continually emerging relationship that supports innovation and programme development, while guiding adaptation to the dynamic realities in complex environments (Patton, 2011). While traditional evaluations generally focus on judging the merit of a programme or improving the programme, DE is a distinct approach that is primarily used for programme development. As sustainability programmes emerge, developmental programme evaluation offers an evaluation approach that is informed by systems thinking and is sensitive to the complex non-linear dynamics present in higher education (Eoyang and Berkas, 1998). DE falls within the larger class of utilization-focused evaluation as the DE process and approach are designed *according to the intended use of the intended users*, instead of for communication with an external audience (Patton, 2008). A developmental approach connects standardized quality systems and theory-driven evaluation with bottom-up participatory evaluation. The developmental evaluator is often a member of the programme development team and responsible for facilitating evaluative discussions, thinking and data-based reflection (Patton, 2011). The organizational form of the developmental evaluator may be an individual or a team of teaching staff and faculty that has included evaluative practices into their work to support ongoing evaluation and learning to inform action (Dick, 2011) that makes an impact in a specific situation (Patton, 2011).

The practice of DE is designed around notions of complex adaptive systems, as characterized by non-linearity, uncertainty, emergence, dynamic interactions, co-evolution and adaptation to changing conditions (Patton, 2011). DE studies the critical incidents and contextual changes that shift programme patterns, move the programme in new directions, and respond to unexpected developments. The evaluation strives to identify and acknowledge sources of uncertainty, while nurturing tolerance for ambiguity. Unlike formative and summative evaluations that aim to improve and test models, a DE must anticipate and expect aspects to emerge and others to decline. This involves searching for anticipated consequences and tracking dynamic interactions among key players (formal and informal, planned and unplanned).

Where certain types of evaluators are oriented to controlled and fixed, standardized interventions, the developmental evaluator expects adaptation while tracking and documenting what, how and why change occurs. The DE co-evolves as new programme processes and new outcomes emerge, requiring new evaluation design elements and measures. Snowden and Boone (2007) argue that in a complex context, in the realm of 'unknown unknowns', leaders must poke around, wait patiently, get a reading, and then determine which emerging patterns will succeed. In higher education sustainability programmes, Snowden and Boone's (2007) approach of probe, sense, respond, can be used by the developmental evaluator to learn by doing and observing to create an adaptive evaluation approach with flexible methods and dynamic designs. With the use of systems-based enquiry, the DE examines the ways in which the various programme settings, components and actors interact with one another, and explores how particular activities and interactions trigger changes in other dynamics of the programme. The systems-based enquiry incorporates systems ideas into the evaluation and facilitates appropriate and relevant systems-sensitive interpretations of the findings grounded in the context of the programme (Patton, 1994).

Continuous learning

Continuous learning is intentionally embedded into the developmental evaluation process. Throughout the DE, the evaluator(s) and the programme developers not only take in information as the programme proceeds, but also commit to integrating the evaluation findings into their thinking and decision-making processes, thus using their learning to refine, adapt and change their approaches as needed (Preskill and Beer, 2012). In contrast to traditional evaluation approaches DE is embedded rather than detached, continuous rather than episodic, and – most importantly – oriented toward individual and organizational learning, not summative judgment (Dozois et al., 2010). DE aims to capture real-time learning, in the midst of uncertainty, to support better understanding of contexts, make data-informed decisions, and guide ongoing adjustments and developments to the programme and its related strategy (Patton, 2011). To direct this learning process, Dozois et al. (2010) recommend developing a learning framework early in the DE process. In comparison to a traditional evaluation framework with established outcomes, targets and indicators, the learning framework is co-created with key stakeholders to guide programme development by mapping key programme challenges and opportunities, highlighting potential areas for learning, and identifying feedback mechanisms for open communication between evaluators and programme developers (Dozois et al., 2010). In addition to supporting the work of programme developers, the learning framework helps the developmental evaluator to be strategic and intentional about where to focus their energy and attention (Dozois et al., 2010), reinforcing the feedback between learning and programme development.

The role of the developmental evaluator

The role of the developmental evaluator combines the responsibility of a traditional evaluator with that of a strategic learning partner and facilitator. Like the traditional evaluator, the developmental evaluator must employ rigorous evaluative and critical thinking to examine and validate the assumptions that underlie the chosen programme strategies. Data collection instruments require adequate testing, sufficient sample sizes, and the triangulation among data gathered from multiple methods and sources according to quantitative and qualitative evaluation standards, which can include both internal and external approaches to evaluation. Throughout the collaborative DE process, the developmental evaluator facilitates stakeholder engagement by framing what the programme is trying to change; how to rigorously collect information; and how to make sense of sometimes very ambiguous data in real time to help programme developers and stakeholders make timely decisions.

Independent of the organizational form, the role of the evaluator facilitates sense-making activities: interpreting, synthesizing and generating insights and recommendations for programme development using multiple forms of written and verbal communications that feed back into the programme design and implementation. A key part of a developmental evaluators' role is to help programme developers and stakeholders surface and test their assumptions, articulate and refine their models, extend their understanding and cultivate a culture that supports learning. These activities help to develop and maintain an adaptive orientation in complex and dynamic environments (Dozois et al., 2010). This approach has much in common with the broader approach to continuous quality improvement and innovation now being used in many higher education institutions around the world.

With a description of the developmental evaluation approach to programmes, we now provide a case example.

Programme-level: applying the developmental evaluation framework

By 2012, Dalhousie's College of Sustainability (CoS) through its Environment, Sustainability, and Society Major (ESS Major), had taught approximately one tenth of the university undergraduate student population (1250 of the 10,000) in a broad range of degree programmes. The ESS classes offered an integration of disciplinary foundations with collaborative and exploratory teaching and learning methods led by interdisciplinary faculty teaching teams. Students were immersed in experiential learning environments developing knowledge, attitudes and skills in multiple literacies, focusing on complexity (systems thinking competence) as well as on interdisciplinarity, self-awareness and engagement (interpersonal competence). After three years of operation, the question was whether the ESS Major was reaching the intended outcomes and vision of the school, faculty, staff and students who had developed this programme.

In response to the need to support real-time evaluative learning in the development of the ESS Major programme and a lack of any comprehensive programme assessment model or tools to measure student learning in areas such as complexity (systems thinking competence) as well as on interdisciplinarity, self-awareness and engagement (interpersonal competence), the CoS hired a research associate in 2012 to design and develop a programme evaluation. The developmental evaluation approach proved to be a natural fit for the evaluation of the ESS programme and was well suited for a pre-formative DE approach. The ESS Major chose this approach because it responds to the call for multilateral relationships among faculty, students, staff and stakeholders for programme development (Van der Leeuw et al., 2012) and because the programme developers continue to work with emerging ideas and visionary hopes. With the aid of developmental evaluation, the ESS programme will probably evolve into a model more fully conceptualized and suitable for formative and summative evaluations.

A collaborative participatory process is at the core of the ESS programme development. The CoS was established as a result of a two-year consultation process that included a two-day university-wide workshop with 100 participants. Building on that initial workshop, the CoS hosts annual ESS curriculum workshops that invite all ESS teaching and cross-appointed faculty, graduate TAs, staff and students. The last three curriculum workshops (in 2010, 2011, 2012) gathered 15–20 students, faculty and staff to explore challenges and possibilities; generated and experimented with new ideas; and innovated the ESS programme model. The following describes a proposed DE for the ESS programme with the aim of discovering and articulating principles to a DE best suited for emerging sustainability programmes in higher education.

Developmental evaluation

The research associate/evaluator began with scoping the evaluation in accordance with ESS guiding principles and key programme elements (Dozois et al., 2010). This action involved a review of programme history, founding documents, curriculum and programme structure to understand the specific context and nature of the ESS Major programme. Next, the evaluator created a flexible learning framework (Dozois et al., 2010) that would employ evaluative findings to map key challenges and opportunities and inform programme development, innovation and experimentation. Table 10.2 provides an overview of the fundamental parts of the ESS programme under review. The elements of the review were developed out of a systems analysis framework (Foster-Fisherman et al., 2007) that demarcates inputs, outputs and outcomes. Tracking changes in outcomes will be a key area of learning as the evaluation

Table 10.2 Preliminary learning framework for the ESS Major programme developmental evaluation

Inputs	Outputs	Outcomes		
		Short-term	*Intermediate-term*	*Long-term*
Learning setting	Reach (# of students and stakeholders)	Enhanced attitudes, perceptions, and/or knowledge	Change in skills: Applied learning to enhance sustainability behaviours	Effectiveness: Improved performance due to enhanced behaviours
Teaching setting	Submitted course work			
Curriculum content	Alumni career path			
Student profile/ background				

adapts to best capture the expected and emerging programme trajectory. This learning framework will act as a living document, to be updated regularly to reflect current learning challenges (Dozois et al., 2010).

After determining the learning framework and the targets (input, output, outcome) of the DE, the next step involved in designing a framework is to describe how these areas of the ESS programme would be evaluated. For the ESS programme evaluation, a mixed methods approach is used to build on existing practices while introducing new forms of enquiry to monitor qualitative shifts. Current evaluation practices include tracking student profiles and recording teaching and learning settings (Table 10.3).

Since its launch, the ESS programme has conducted annual student online surveys to explore the academic profiles of ESS students and the influence of the ESS programme on student decisions to attend and stay at Dalhousie. All ESS classes participate each term in university-wide standardized Student Rating of Instructor surveys that assess students' opinions of their course instructors. Students in ESS classes have supplemental questions that ask for their perceptions of the ESS team-teaching approach; lectures and tutorials; lecture slides and audio-visual material; Online Web Learning resources; and class content. Lastly, written comment questions offer students the opportunity to describe the positive aspects and improvement areas. These surveys, along with student profiles and teaching and learning records, help develop an understanding of the fundamental programme components, including operations, class resources, academic regulations and programme norms.

The proposed programme DE will focus on understanding how the design and implementation of the ESS teaching and learning approaches

Table 10.3 Current evaluation practices of ESS Major programme focusing on factual data

Evidential/ Evocative enquiry	Primary approach			
	Collect what?	From whom?	How?	When?
Facts	Student profile	Registrar office, students	Registration, entry survey	Annually
	Learning & teaching setting	Professor, TAs	Curriculum (environment, pedagogical approaches, activities, assignments, formats of assessments)	Annually
	Curriculum	Professor	Curriculum map (syllabi, content, learning outcomes)	Annually

impacts learning outcomes. The teaching faculty of the ESS programme worked with a subset of the five sustainability key competences as learning outcomes. This subset includes: systems-thinking (complex systems) and interpersonal competence (multiple literacies, interdisciplinarity, self-awareness, and engagement). Without any comprehensive quantitative instruments to measure learning in these areas, the DE design requires an exploratory approach combining various qualitative enquiries to track changes in students' knowledge, attitudes, skills and behaviours (Table 10.4).

The use of markers, observation, reflections and stories provides multiple perspectives on the impacts of the ESS programme and the influence on students' learning relative to the five key sustainability competences. The self-efficacy scale will allow students to self-assess progress related to key competences. The self-efficacy survey will be conducted online at the beginning, midway, and post completion of the ESS Major.

The developmental evaluator will conduct direct observation of a selection of programme components, including lectures, seminars, tutorials and presentations. This information will supplement curriculum documents that describe intended teaching and learning settings. These observations allow collecting immediate feedback on programme

Table 10.4 Proposed evaluation components of the new ESS enquiry framework based on qualitative aspects

Evidential/ Evocative enquiry	Primary approach			
	Collect what?	*From whom?*	*How?*	*When?*
Markers	Level of self-efficacy	Students	Self-efficacy survey of sustainability competences	3x/programme: pre, during, post
Observation	Teaching settings displayed in classroom that convey sustainability competences	Developmental evaluator	Direct observation	Periodically
Reflections	New learning, 'aha! moments' related to sustainability competences	Students	Online journals	Periodically
Stories	Most significant change story	Teaching assistants, faculty, students, alumni	Focus groups using appreciative enquiry	Bi-annually (Term-end)

approaches and activities vis-à-vis attainment of sustainability key competences.

At present, ESS students are required to submit reflective journal entries as part of some ESS course requirements; in the future the DE plans to adjust course requirements such that all ESS courses will include a self-reflection component that encourages students to track their learning progress (knowledge, attitudes, values, skills) throughout the programme. These reflective submissions are one way to capture programme outcomes by analysing students' ideas, personal thoughts, experiences and insights throughout the learning process.

Collecting stories provides information concerning unexpected outcomes, programme performance based on the best success stories, and a form of dynamic values-enquiry (Dart and Davies, 2003). During the sense-making stage of the evaluation, the programme developers and developmental evaluator will search for significant programme outcomes from the collected stories and then deliberate on the value of these outcomes. This process will facilitate programme development by focusing the programme towards explicitly valued directions.

The above forms of enquiry are new to the ESS programme and are open to adaptation. The proposed DE is a starting point; further measures and tracking mechanisms will develop quickly as outcomes emerge. The ESS Major has a clear sense of the intended direction of the programme, with established guiding principles and distinct programme features. The DE will help to develop, innovate and articulate the programme strategy and its progress markers.

Continuous learning

While the quality of data collection and design is crucial, the ultimate criterion for the success of the DE approach is the extent to which the information generated through data collection and reflections are used to further innovate the ESS programme and adapt the DE design. To ensure the DE works well at the college, the developmental evaluator depends on the support of programme leadership to help foster a commitment to learning and change among teaching faculty and staff. The ESS programme will include an evaluation workshop at the annual curriculum retreat to review and analyse the collection of data from the previous school year. The workshop will encourage an organizational culture of information sharing, collaboration, learning from failure or mistakes, and using data for decision-making. Faculty, staff and students will gather to share multiple understandings to co-generate knowledge (Blackstock et al., 2007; Van der Leeuw et al., 2012). In addition to the annual workshop, regular collaborations between the evaluator and individual teaching teams may be useful to review timely feedback and facilitate ongoing learning to inform strategic decisions for the ESS programme and the evaluation approaches.

Role of the developmental evaluator

The role of the ESS developmental evaluator is currently being filled through a contracted research assistant. Although the college will have a designated developmental evaluator, s/he will be a part of a team of staff and teaching faculty who collaborate to conceptualize, design and test new approaches in a long-term process of continuous improvement. The developmental evaluator's primary function in the team will be to elucidate team discussions with evaluative questions, data and logic, and facilitate data-based decision-making in the developmental process (Patton, 1994).

Discussion

Course-level challenges

Problem- and project-based learning (PPBL) approaches are used to structure individual courses or entire curricula. However, using a PPBL approach for courses without a progressive model of incorporating PPBL across the curriculum can be defeating (Sterling and Thomas, 2006; Brundiers et al., 2010). Students and instructors have to switch within the same semester from attending a majority of classes designed in the regular mode (instructors as knowledge providers and students as knowledge receivers) to a few classes designed in PPBL mode, which turns roles upside down and often expects everyone to succeed without adequate preparation or support (Savery, 2006). Sustainability programmes where PPBL is not the modus operandi thus put a lot of pressure on individual courses. These pressures create a need to introduce students to new learning techniques, offer opportunities to apply them, and enable reflection and evaluation of process and outputs. As many of the PPBL courses also involve project clients and stakeholders, additional pressures arise from the need to balance educational objectives with project objectives for products that are relevant to clients. We discuss four approaches to advancing PPBL courses for sustainability.

One way to address the challenge is to progressively incorporate PPBL across the curriculum. Brundiers et al. (2010) have outlined a model for how to prepare students from the outset to participate successfully in PPBL courses. The model incorporates PPBL components strategically into courses along the entire programme and calls for an accompanying, required introductory course to allow students to practise key elements of PPBL (e.g. teamwork, project management, self-directed learning and stakeholder engagement) as well as of solution-oriented research methodologies.

A second way to address the challenge is to alleviate the various pressures related to any specific PPBL courses. These pressures often arise from time constraints. To circumvent this challenge, course instructors and project clients could negotiate a series of cumulative, semester-long PPBL courses and internship experiences designed to better balance educational and project objectives. This comprehensive and integrated approach of a series of cumulative courses reduces pressure to develop actionable solution options to sustainability issues within the artificial confines of the university calendar, which is structured along semesters.

A third way to address the challenge is to encourage instructors to share the significant role of instructional facilitator. In the 'Don't Landfill it!' course case presented earlier, we combined graduate and undergraduate students in teams and assigned graduate students the role of team coaches (peer-mentoring). Furthermore, steps could be taken to recruit *subject matter experts* from the professional and academic community to support instructors in ensuring quality of any content directly delivered to students.

Finally, Brundiers et al. (2013) have highlighted the role of a Transacademic Interface Manager (TIM). A TIM should assist across all phases of the project by helping instructors to set up projects, develop partnerships with project clients, organize the PPBL course itself, facilitate collaboration and project management, and coach students on professional and social skill development. The TIM is typically a staff member funded through the university to assist faculty in these various efforts.

Programme-level challenges

To date there are no 'universally' agreed upon, standardized learning outcomes for sustainability science education. Developmental evaluations, by collecting evidence as well as forming experience and tacit knowledge, will probably serve as a key foundation for developing formative and summative evaluations in sustainability science education. We identified three areas that warrant attention when it comes to adapting developmental evaluations for sustainability programmes to the dynamic, complex context of sustainability challenges: the frame of reference of the evaluation, the link to course evaluations, and the involvement of stakeholders.

For sustainability education programmes, developmental evaluation is adaptive within a normative frame of reference. This frame of reference is defined by the sustainability principles guiding the development of sustainable solution options (cf. Gibson, 2006) and by sustainability problem-solving methodologies (cf. Wiek et al., 2012). The challenge – for Dalhousie's CoS and other sustainability programmes – is to operationalize both sustainability principles and sustainability key competences into measurable programme-level learning outcomes. The CoS emphasizes learning outcomes related to knowledge and skills as well as attitudes and values. While this commitment to an expanded set of learning outcomes reflects new thinking in educational and sustainability sciences (cf. Sipos et al., 2008), it is remarkable considering the often implicit, yet widely held belief of 'value-free' and 'objective' science. The challenge will be to find a compromise between formulating values and attitudes that are neither overly prescriptive nor meaninglessly vague.

The second area relates to course evaluations as inputs into programme evaluation (cf. Table 10.2). Summative course evaluations, where students evaluate the course and instructors' performance, need to be complemented with formative and summative programme evaluations that engage students in assessing overall learning gains and push students to explicate what contributed to their learning and why (Seymour et al., 2000; Eyler, 2002). The combination of these two types of evaluation might better hold instructors *and* students accountable for learning success and reinforces the responsibility for a two-way learning relationship.

The third area relates to involving stakeholders in developmental evaluation. To meet calls in sustainability science for extended peer review and quality assurance (Pereira and Funtowicz, 2005; Brundiers and Wiek, 2011; Talwar et al., 2011), community partners need to be included in programme evaluation. This reality is especially acute as course-level changes (cf. those discussed in section 'Course-level: applying the framework for PPBL courses') make community partners more integral to individual course design process in sustainability than in other fields.

Conclusion

This chapter presented insights into how undergraduate sustainability programmes translate into practices proposed in the pertinent areas of literature, for example sustainability science, transdisciplinarity and education (Barth and Michelsen, 2013). These practices draw increasingly on problem- and project-based learning approaches to course design. Hence, we presented an analytical-evaluative framework for designing, implementing and evaluating problem- and project-based sustainability courses. We presented the framework of developmental evaluation as an approach to evaluate entire sustainability programmes. This framework offers a learning-based and adaptive approach to programme evaluation, which is beneficial as many sustainability programmes are still emerging. Both frameworks were applied to appraise two specific cases. Lastly, we presented the concept of sustainability key competences as a guiding framework to formulate and evaluate learning outcomes on course and programme levels. However, more substantive research is needed to address the question of how to operationalize sustainability key competences as salient, measurable outcomes. We hope that our chapter contributes to setting the stage for exploring this and other questions in an intercultural and transdisciplinary manner committed to mutual, multilateral learning for sustainability.

Acknowledgements

We would like to thank the anonymous reviewer as well as the book editors for helpful comments on an earlier version of this article, Aaron Redman for his reflections on the 'Don't Landfill it!' course, which he co-taught, and Michael Bernstein for editorial support.

Notes

1. The term 'experts' includes people with 'officially' recognized expertise in their respective area of specialization, for example, pundits, scientists and CEOs as well as people who hold local or otherwise specific everyday knowledge that is relevant to understanding sustainability problems and possible solution options.
2. Environment, Society and Sustainability Major at the College of Sustainability at Dalhousie University, Canada: http://www.dal.ca/faculty/sustainability.html; Sustainability programme at the School of Sustainability at Arizona State University, USA: http://schoolofsustainability.asu.edu; Leuphana University, Germany: http://www.leuphana.de/college/bachelor/studiengang-major/environmental-and-sustainability-studies.html
3. Numbers are averages of the three programmes for the year 2011.
4. The terms 'assessment' and 'evaluation' are often synonymously used. We follow definitions provided by Marzano and Kendall (2006) who define assessment as gathering information about student achievement, based on students' outputs; evaluation as judging on students' learning success, that is, to determine to what extent learning objectives have been achieved and why; and measurement as assigning marks, as basis for grades.
5. Prior to the research, the authors had obtained approval for their research, involving human subjects, from the Arizona State University's Institutional Review Board.

References

Ash S. L., Clayton P. H. (2004) 'The Articulated Learning: An Approach to Guided Reflection and Assessment'. *Innovative Higher Education* 29 (2) 137–154.

Bammer G. (2005) 'Integration and Implementation Sciences: Building a New Specialization'. *Ecology and Society*, 10(2)6.

Barth M., Godemann J., Rieckmann M., Stoltenberg U. (2007) 'Developing Key Competencies for Sustainable Development in Higher Education'. *International Journal of Sustainability in Higher Education*, 8 (4) 416–430.

Barth M., Michelsen G. (2013) 'Learning for Change: An Educational Contribution to Sustainability Science'. *Sustainability science*, 8 (1) 103–119.

Bereiter C., Scardamalia M. (2003) 'Learning to Work Creatively'. In E. De Corte, L. Verschaffel, N. Entwistle, J. van Merrienboer (eds) *Powerful Learning Environments: Unravelling Basic Components and Dimensions*, 1st edn (Amsterdam: Pergamon), 55–68.

Biggs J. (1999a) 'What the Student Does: Teaching for Enhanced Learning'. *Higher Education Research & Development*, 18 (1) 57–75.

Biggs J. (1999b) *Teaching for Quality Learning at University* (Buckingham, UK: Society for Research Into Higher Education and Open University Press).

Blackstock K., Kelly G., Horsey B. (2007) 'Developing and Applying a Framework to Evaluate Participatory Research for Sustainability'. *Ecological Economics*, 60, 726–742.

Blumenfeld P. C., Soloway E., Marx R. W., Krajcik J. S., Guzdial M., Palinscar A. (1991) 'Moti- Vating Project-Based Learning: Sustaining the Doing, Supporting the Learning'. *Educational Psychologist*, 26 (3/4) 369–398.

Bornman G. M. (2004) 'Programme Review Guidelines for Quality Assurance in Higher Education: A South African Perspective'. *International Journal of Sustainability in Higher Education*, 5 (4) 372–383.

Brundiers K., Wiek A. (2011) 'Educating Students in Real-world Sustainability Research: Vision and Implementation'. *Innovative Higher Education*, 36 (2) 107–124.

Brundiers K., Wiek A. (2013) 'Do We Teach What We Preach? An International Comparison of Problem- and Project-based Learning Courses in Sustainability'. *Sustainability*, 5 (4) 1725–1746.

Brundiers K., Wiek A., Kay B. (2013) 'The Role of Transacademic Interface Managers in Transformational Sustainability Research and Education'. *Sustainability*, 5, 4614–4636.

Brundiers K., Wiek A., Redman C. L. (2010) 'Real-world Learning Opportunities in Sustainability: From Classroom into the Real World'. *International Journal of Sustainability in Higher Education*, 11 (4) 308–324.

Cash D. W., Clark W. C., Alcock F., Dickson N. M., Eckley N., Guston D. H., Jaeger J., Mitchell R. B. (2003) 'Knowledge Systems for Sustainable Development'. *Proceedings of the National Academy of Sciences*, 100 (14) 8086–8091.

Crow M. (2010) 'Organizing Teaching and Research to Address the Grand Challenges of Sustainable Development'. *BioScience*, 60 (7) 488–489.

Dart J., Davies R. (2003) 'A Dialogical, Story-Based Evaluation Tool: The Most Significant Change Technique'. *American Journal of Evaluation*, 24 (137) 137–155.

de Haan G (2006) 'The BLK "21" Programme in Germany: A "Gestaltungskompetenz"-based Model for Education for Sustainable Development'. *Environmental Education Resources*, 1, 19–32.

Dick B. (2011) 'Action Research Literature 2008–2010: Themes and Trends'. *Action Research*, 9(2) 122–142.

Donnelly R., Fitzmaurice M. (2005) 'Collaborative Project-based Learning and Problem-based Learning in Higher Education: Consideration of Tutor and Student Roles in Learner-focused Strategies'. In G. O'Neill, S. Moore, B. McMulling (eds) *Emerging Issues in the Practice of University Learning and Teaching* (Dublin: All Ireland Society for Higher Education (AISHE)), 87–98.

Dozois E., Langlois M., Blanchet-Cohen N. (2010) *DE 201: A Practitioner's Guide to Developmental Evaluation* (The J. W. McConnell Family Foundation and The International Institute for Child Rights and Development. Library and Archives Canada Cataloguing in Publication).

Eoyang G., Berkas T. (1998) 'Evaluating Performance in a Complex Adaptive System'. In M. Lissack, H. Gunz (eds) *Managing Complexity in Organizations* (Westport, CT: Quorum Books).

Eyler, J. (2001) 'Creating your Reflection Map'. *New Directions for Higher Education*, 2001, (114) 35–43.

Ferrer-Balas D., Adachi J., Banas S., Davidson C. I., Hoshikoshi A., Mishra A., Motododa Y., Onga M., Ostwald M. (2008) 'An International Comparative Analysis of Sustainability Transformation across Seven Universities'. *International Journal of Sustainability in Higher Education*, 9 (3) 295–316.

Foster-Fisherman P. G., Nowell B., Yang H. (2007) 'Putting the System Back into Systems Change: A Framework for Understanding and Changing Organizational and Community Systems'. *American Journal of Community Psychology*, 39, 197–216.

Fraser M. W., Galinsky M. J. (2010) 'Steps in Intervention Research: Designing and Developing Social Programs'. *Research on Social Work Practice*, 20(5) 459–466.

Frisk E., Larson K. L. (2011) 'Educating for Sustainability: Competencies & Practices for Transformative Action'. *Journal of Sustainability Education*, (2) http://www.jsedimensions.org/wordpress/content/educating-for-sustainability-competencies-practices-for-transformative-action_2011_03/

Gibson R. B. (2006) 'Sustainability Assessment : Basic Components of a Practical Approach'. *Impact Assessment and Project Appraisal*, 24(3) 170–182.

Gruenewald D. A. (2003) 'The Best of Both Worlds: A Critical Pedagogy of Place'. *Educational Researcher*, 32 (4) 3–12.

Hmelo-Silver C. E. (2004) 'Problem-Based Learning: What and How Do Students Learn?' *Educational Psychology Review*, 16 (3) 235–266.

Jones P., Selby D., Sterling St (eds) (2010) *Sustainability Education: Perspectives and Practice Across Higher Education* (London and Washington DC: Earthscan).

Knight L. A. (2006) 'Using Rubrics to Assess Information Literacy'. *Reference Services Review*, 34 (1) 43–55.

Marzano R. J., Kendall J. S. (2006) *The New Taxonomy of Educational Objectives* (Thousand Oaks, CA: Corwin Press, A SAGE Publications Company).

Meehan B., Thomas I. (2008) 'Research and Solutions: University Education for Sustainability: Projects, Problems, and Professionalism'. *Sustainability: The Journal of Record*, 1(2) 124–129.

Moore J. (2005) 'Seven Recommendations for Creating Sustainability Education at the University Level: A Guide for Change Agents'. *International Journal of Sustainability in Higher Education*, 6 (4) 326–339.

Moust J. H., Berkel H. J. V., Schmidt H. G. (2005) 'Signs of Erosion: Reflections on Three Decades of Problem-based Learning at Maastricht University'. *Higher Education*, 50 (4) 665–683.

Nicol D. J., Macfarlane-Dick D. (2006) 'Formative Assessment and Self-regulated Learning: A Model and Seven Principles of Good Feedback Practice'. *Studies in Higher Education*, 31 (2) 199–218.

Patton M. Q. (1994) 'Developmental Evaluation'. *Evaluation Practice*, 15 (3) 311–319.

Patton M. Q. (2008) *Utilization-Focused Evaluation*, 4th edn (Saint Paul, MN: SAGE Publications Inc.)

Patton M. Q. (2011) *Developmental Evaluation: Applying Complexity Concepts to Enhance Innovation and Use* (New York: The Guildford Press).

Pereira A. G., Funtowicz S. (2005) 'Quality Assurance by Extended Peer Review: Tools to Inform Debates, Dialogues, and Deliberations'. *Technikfolgenabschätzung Theorie und Praxis*, 14 (2) 74–79.

Potvin L., Cargo M., McComber A. M., Delormier T., Macaulay A. C. (2003) 'Implementing Participatory Intervention and Research in Communities:

Lessons from the Kahnawake Schools Diabetes Prevention Project in Canada'. *Social Science & Medicine*, 56 (6) 1295–1305.

Preskill H., Beer T. (2012) *Evaluating Social Innovation* (FSG and Center for Evaluation Innovation).

Roessingh H., Chambers W. (2011) 'Project-Based Learning and Pedagogy in Teacher Preparation: Staking Out the Theoretical Mid-Ground'. *International Journal of Teaching and Learning in Higher Education*, 23 (1) 60–71.

Rowe D. (2007) 'Education for a Sustainable Future'. *Science*, 317 (5836) 323.

Savery J. R. (2006) 'Overview of Problem-based Learning: Definitions and Distinctions', *Interdisciplinary Journal of Problem-based Learning*, 1 (1) 9–20.

Segalas J., Ferrer-Balas D., Svanström M., Lundqvist U., Mulder K. F. (2009) 'What Has to be Learnt for Sustainability? A Comparison of Bachelor Engineering Education Competences at Three European Universities'. *Sustainability Science*, 4 (1) 17–27.

Seymour, E., Wiese, D. J., Hunter, A., Daffinrud, S. M. (2000, March) 'Creating a Better Mousetrap: On-Line Student Assessment of Their Learning Gains'. In *National Meeting of the American Chemical Society*.

Shriberg M. (2002) 'Institutional Assessment Tools for Sustainability in Higher Education: Strengths, Weaknesses, and Implications for Practice and Theory'. *International Journal of Sustainability in Higher Education*, 3 (3) 254–270.

Sipos Y., Battisti B., Grimm K. (2008) 'Achieving Transformative Sustainability Learning: Engaging Head, Hands and Heart'. *International Journal of Sustainability in Higher Education*, 9 (1) 68–86.

Snowden D. J., Boone M. E. (2007) 'A Leader's Framework for Decision Making'. *Harvard Business Review*, 85 (11) 1–9.

Spangenberg J. H. (2011) 'Sustainability Science: A Review, an Analysis and Some Empirical Lessons'. *Environmental Conservation*, 38, 275–287.

Stauffacher M., Walter A. I., Lang D. J., Wiek A., Scholz R. W. (2006) 'Learning to Research Environmental Problems from a Functional Socio-cultural Constructivism Perspective: The Transdisciplinary Case Study Approach'. *International Journal of Sustainability in Higher Education*, 7 (3) 252–275.

Sterling S., Thomas I. (2006) 'Education for Sustainability: The Role of Capabilities in Guiding University Curricula'. *International Journal of Innovative Sustainable Development*, 1 (4) 349–370.

Svanström M., Lozano-García F. J., Rowe D. (2008) 'Learning Outcomes for Sustainable Development in Higher Education'. *International Journal of Sustainability in Higher Education*, 9 (3) 339–351.

Talwar S., Wiek A., Robinson J. (2011) 'User Engagement in Sustainability Research'. *Science and Public Policy*, 38 (5) 379–390.

Thomas I. (2009) 'Critical Thinking, Transformative Learning, Sustainable Education, and Problem-based Learning in Universities'. *Journal of Transformative Education*, 7 (3) 245–264.

Van der Leeuw S., Wiek A., Harlow J., Buizer J. (2012) 'How Much Time Do We Have? Urgency and Rhetoric in Sustainability Science'. *Sustainability Science*, 7 (S1) 115–120.

Wiek A., Lang D. J. (2011) 'Transformational Research for Sustainability: A Practical Methodology', School of Sustainability, Arizona State University.

Wiek A., Ness B., Schweizer-Ries P., Brand F. S., Farioli F. (2012) 'From Complex Systems Analysis to Transformational Change: A Comparative Appraisal of Sustainability Science Projects'. *Sustainability Science* 7(Suppl.) 1–20.

Wiek A., Withycombe L., Redman C. L. (2011) 'Key Competencies in Sustainability: A Reference Framework for Academic Program Development'. *Sustainability Science*, 6, 203–218.

Wiek A., Xiong A., Brundiers K., van der Leeuw S. (in press) 'Integrating Problem- and Project-Based Learning into Sustainability Programs: A Case Study on the School of Sustainability at Arizona State University'. *International Journal of Sustainability in Higher Education*.

Wiggins G., McTighe J. (1998) *Understanding by Design* (Association for Supervision and Curriculum Development (ASCD)).

11
Quality Management of Education for Sustainability in Higher Education

Geoff Scott

Overview

There is growing interest in building education for social, cultural, economic and environmental sustainability (EfS) into the core activities of universities and colleges around the world, not only into their research and engagement activities and their operations but also, most importantly, into their curriculum. However, having a well-designed EfS programme approved or a strategic plan for the area in place does not ensure that it will be effectively and productively put into practice, continuously enhanced or sustained.

Taking good ideas and making them happen – consistently and well – is a key challenge for higher education institutions around the world. This chapter discusses how this can be done in the distinctive area of EfS by bringing together research on the area, experience from two decades' work on quality development and innovation and higher education around the world and a practical case study.

The chapter takes a quality management system for learning and teaching (L&T) that has led to significant improvement in the educational programmes in one Australian university and outlines how it can apply to the specific area of EfS. It also briefly outlines how staff have been successfully and sustainably engaged with using this system, how desired L&T changes can be consistently enhanced and sustainably implemented, and the most effective approaches to change leadership of EfS in our colleges and universities.

What follows is based on a keynote presentation on improving learning and teaching quality in higher education given at the 2011

South African Association of Institutional Research national conference which was later published in the *South African Journal of Higher Education* (Scott, 2013).

Context

This is a turnaround moment for higher education world wide – the traditional 19th-century fixed-timetable, content-focused, institutionally delivered and single-discipline-based model of higher education is coming under increasing scrutiny as:

- access to higher education is widened and institutions are confronted with the dilemmas of how best to balance growth with quality, access with excellence, mission with market;
- universities and colleges are subjected to funding cuts and must manage growth, costs and risk in an environment of growing regulation and financial constraint;
- student expectations change and they increasingly seek just-in-time support, real-world learning and placements, targeted learning assistance, convenient access and value-for-money in their studies, along with successful employment or further study outcomes;
- the IT revolution reshapes the world of information and interaction, with the consequence that higher education institutions are no longer seen as being the sole repository of key up-to-date information and solutions;
- governments are confronted with having to respond appropriately to the challenges of increasing globalization, educational competition, fractious division, budget shortfalls and the impact of a rapidly unfolding climate crisis, along with having to figure out how best to manage the challenges of social, cultural, economic and environmental sustainability.

Interlaced with these broader change forces is a growing movement which expects institutions of higher education to engage their staff, students and stakeholders with greater consistency in creating more sustainable futures:

- The period 2005–2014 marks the UN Decade of Education for Sustainable Development and global efforts to integrate sustainability more consistently into higher, further and informal adult education.

- In June 2012 the UN Conference on Sustainable Development (Rio+20) was held in Brazil. The Rio+20 commitments call for universities to become models of best practice and transformation.
- A formal Higher Education Treaty for Rio+20 has been generated through a collaborative process involving key international and national agencies, associations and organizations. The document, which is an official Rio+20 Treaty, commits the sector to transformational change for sustainability.
- Governments are expecting their universities and colleges to respond to the sustainability imperative. In Australia, for example, the *2009 National Action Plan for ESD* provides a framework for higher education leadership in this area; and peak groups like Universities Australia have given *commitment* to action on it.
- The United Nations, via its UN University, has endorsed some 130 Regional Centres of Expertise in Education for Sustainable Development around the world. This is supported by the UNECE ESD Competences framework which guides development across the higher education curricula.

However, as already noted, a university or college having a good idea on how to respond to this agenda or signing off on a commitment to act in support of sustainable development does not make it happen. And it is to this challenge and to a proven way to ensure effective change management, implementation, quality assurance and improvement for desired EfS futures in our institutions of higher education that this chapter turns.

Running through the strategies discussed in the chapter is the process of evaluation – which is defined as 'the process of making judgements about the worth of what is proposed, how well it is being put into practice and the quality of its impact'. At the heart of the word 'evaluation' is the word 'value': the conclusion by individuals that what is happening, for example in an EfS initiative, is 'good' or 'bad'. There is, therefore, a profound difference between 'change' (something becoming or being made different) and 'progress' (a conclusion by an individual or group that this is of benefit). It is in this way that judgement, personal values and seeking to engage everyone in a university with the call for higher education to take a more central role in addressing the integrated challenges of social, cultural, economic and environmental change poses such a challenge. As George Bernard Shaw is said to have observed: 'Reformers have the misplaced notion that change is achieved by brute logic'.

The University of Western Sydney as a case study

About the University of Western Sydney (UWS)

UWS serves one of the fastest growing regions in Australia with the third-largest economy in the country. The Greater Western Sydney (GWS) region is home to 1.9 million people. It has more indigenous residents than the states of Victoria and South Australia; people from more than 170 countries speaking over 100 different languages live there. The majority of new immigrants to Australia (60%) settle in GWS, with 50 per cent of these arrivals coming from countries like Iraq and Sudan. The region has a high proportion of low-income families and it has the lowest high-school retention rates in the Sydney metropolitan area.

The four key sustainability challenges facing this rapidly developing peri-urban region are: transitioning to a low carbon economy; developing sustainable communities, including sustainable health, transport and housing; ensuring agricultural sustainability and food security; and conserving biodiversity and river health.

UWS has some 38,000 students, more than 50 per cent of whom are the first in their family to attend university. It has five major campuses spanning some 2400 km^2 of the region, along with a pathways college (the UWS College), which is co-located with two high schools and a vocational education and training institution on a sixth campus. A community television station, Television Sydney, is one of its entities. It offers courses ranging from law, medicine, nursing and the biomedical, natural and health sciences to education, engineering, business, the arts, humanities, languages, social sciences, and media and communications. Its key research centres and institutes focus on key issues of social, cultural, economic and environmental sustainability as they play out in a context like GWS and regions similar to it around the world.

With this profile, UWS is, in many ways, typical of a 21st-century metropolitan university. It is an institution focused on addressing the key issues of social, economic and environmental sustainability in its region in a two-way partnership with its community; providing access to a high-quality educational experience for talented people hitherto excluded from tertiary education; ensuring that they graduate and, through this, seeing that their life opportunities and those of their families and the region are profoundly improved.

Central to ensuring that all students who attend UWS receive as high a quality education as possible, are retained, graduate and go onto successful careers and service has been the constant monitoring and addressing of shortfalls in educational performance using a range of key L&T indicators and associated quality processes.

Performance trends in learning and teaching at the UWS

Over the period from 2005 to 2010 UWS saw an improvement across a range of relevant L&T performance indicators:

- The percentage of first preferences from school-leavers wishing to study at UWS increased from 48 per cent to 50.5 per cent.
- Overall satisfaction ratings given in Australia's university-course experience questionnaire increased by 25 per cent.
- Retention increased by 4 per cent.
- Employability grew to be at the national average (81%) and the percentage of UWS students going on to further study exceeded the national average by more than 5 per cent.
- In respect of national awards for L&T, UWS achieved 12 teaching-excellence awards in 2011 from the Australian Learning and Teaching Council, including Australian Teacher of the Year, compared with none in 2005. It also received the Australian Teacher of the Year Award in 2012.
- The UWS Tracking and Improvement System for L&T (TILT) is cited on the good-practice database of the Australian Universities Quality Agency (AUQA).
- In 2008, UWS was the university selected to write the *research and analysis report on student satisfaction and engagement* in higher education for the Review of Australian Higher Education (the so-called Bradley Review).

How this improvement has been achieved

The improvements in L&T performance identified above have been achieved by applying the key lessons of effective change management in higher education as outlined in books like *Turnaround Leadership for Higher Education* (Fullan and Scott, 2009) and *Change Matters: Making a Difference in Education and Training* (Scott, 1999). Specifically, the improvement strategy gives focus to:

- bringing together the right combination of *what* needs to be done with the key lessons from research on *how* to make sure that these desired changes are implemented – consistently and well;

and to:

- building a change-capable culture, where 'culture' simply means 'how we do things around here'.

A Pro Vice-Chancellor (Quality) was appointed in 2004 to help lead the quality-improvement process for the area. This was in recognition of the fact that change does not just happen but that it must be led, and deftly.

In the rest of the chapter first the process and organizing framework used to determine *what* is best given focus in L&T improvement efforts is outlined; then *how* these desired changes have been implemented is discussed and, finally, the strategies for developing a more change-capable and resilient culture are outlined. Our recent work on applying these lessons to EfS indicates that this overall approach can be readily adapted to foster effective innovation, implementation and ongoing improvement specific to the area of education for sustainable development.

The UWS academic standards and quality framework

UWS selected 'Assuring and improving academic standards for learning, teaching and assessment' as one of two themes for its 2011 audit by the Australian Universities Quality Agency (AUQA). (The other theme was 'Quality management for student transition and retention', with a particular focus on a range of equity groups.)

In order to achieve the improvements in L&T performance noted above, it was necessary that UWS, in close collaboration with its staff, develop an overall framework on what assuring and improving academic standards for L&T means in practical terms at the university (see Figure 11.1). This framework applies well to EfS innovations and programmes in particular as well as to learning programmes more generally.

Our commissioned research and analysis report on the area for the *Review of Australian Higher Education* (Scott, 2008), our study of some 500,000 best-aspect and needs-improvement comments from UWS students using the CEQuery qualitative analysis tool, extensive institutional research and targeted benchmarking with like universities within and beyond Australia helped to ensure this framework was comprehensive and valid.

The framework has two components. The top half identifies *what* UWS focuses on in seeking to assure academic standards and quality for learning, teaching and assessment. The lower half represents *how* the university goes about ensuring that these standards are applied, tracked and improved consistently, systematically and effectively. It is the careful and consistent attention to both areas that is seen by the UWS as being necessary to provide for effective academic quality management, improvement, implementation and assurance.

The top section of the UWS framework has the four interlocking domains, each with its own set of standards and quality checkpoints:

1. course design;
2. learning support;
3. delivery;
4. learning outcomes and assessment (impact).

Whereas Domains 1 and 2 in Figure 11.1 are concerned with assuring the quality and standards of *inputs*, Domains 3 and 4 focus more on the

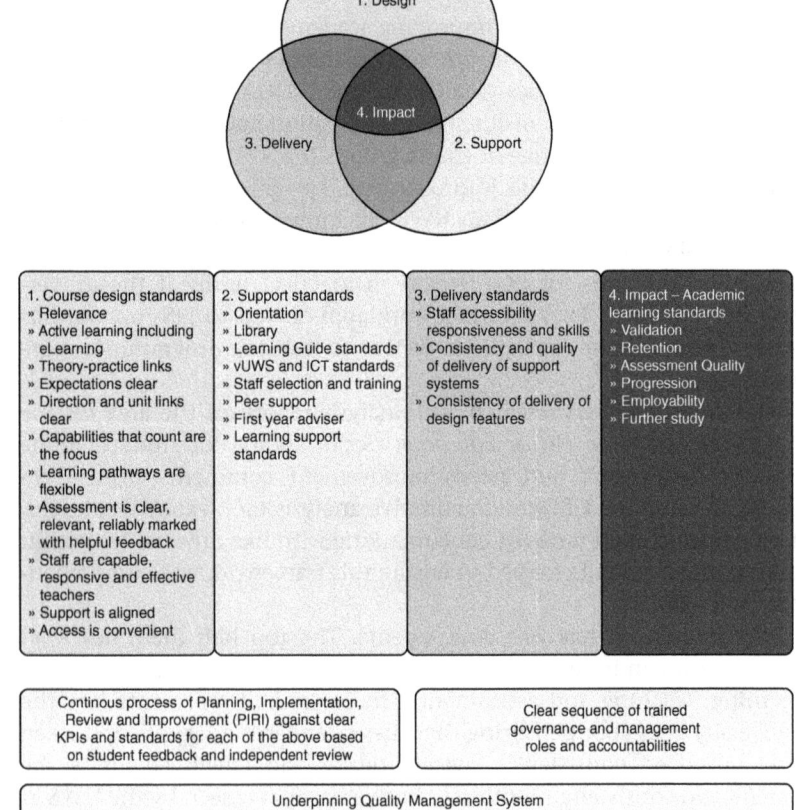

Figure 11.1 UWS framework for assuring academic standards and quality in L&T

quality of *outcomes*. UWS sees that the key test of L&T quality and standards resides in the fourth domain (achieving a demonstrably positive impact on the capabilities that count for its graduates to have productive careers and to contribute to a socially just and sustainable future).

This framework distinguishes between (3) teaching (what teachers do) and (4) learning (what learners do). In this context, 'learning' concerns the extent to which the capabilities and competences of students have developed in a desirable (professionally and socially relevant) direction over the course of their studies.

Because of this, giving focus to validating course-level learning outcomes (4) using a suite of agreed external and institutional quality assurance reference points is a key *first* step in course development. The university seeks to design engaging courses (1) against these validated learning outcomes to build student achievement progressively while ensuring that there is aligned support (2) and consistent delivery (3). In doing this, the university aims to ensure that design, support and delivery decisions are not only relevant but also aligned, mutually reinforcing, outcomes-focused and evidence-based.

Every component of the framework in Figure 11.1 comprises those empirically determined quality checkpoints known to optimize the retention and engagement of students in productive learning. Specific focus is given to ensuring that every staff member – academic or professional – can see what contribution their role makes in helping to retain and engage UWS students and that they are acknowledged for excellence in that work.

The university recognizes that what matters to students is the combined and consistent quality of all four domains in Figure 11.1 (that is, the total university experience) and that it is the extent to which the valid standards in all four areas are delivered, monitored and improved effectively and consistently that determines the quality of graduates.

When new programmes are designed using the framework, concurrent attention is given to all four standard domains. For example, as courses, including those concerned with EfS, are designed (Domain 1 in Figure 11.1), there is close attention to ensuring that they are relevant to real-world practice; give focus to problem-based and integrated cases from the field; ensure active learning; foster team-based work; show clear and consistent links between practice and the underpinning theory; make clear that what is being given focus is the development of the capabilities that will count subsequently for productive professional and societal action; that they use experienced staff who are responsive, competent in the area taught and are effective teachers;

and that all of the learning experiences built into each unit of study enable students to successfully address integrated, problem-based assessment tasks.

Of particular relevance in the area of EfS is the systematic use of the campus as a living laboratory for both formal and informal learning and research about social, economic and environmental sustainability. This 'real-world' approach to university learning is in recognition of Bill Spohn's (2003) observation that people are more likely to act their way into new ways of thinking than think their way into new ways of acting. This, in turn, echoes the old Chinese proverb: 'I hear and I forget, I see and I remember, I do and I understand' along with John Dewey's (1933) early-20th-century approaches to active, real-world learning and reflection. In a similar fashion it is possible to use the surrounding community as both a site and source for EfS research and learning. For example, UWS is using the initiatives of the UNU-endorsed RCE-GWS which it hosts to provide real-world learning opportunities and field placements for our students along with those from surrounding schools and vocational education colleges. The flagship of this initiative is the UWS Riverfarm living learning laboratory.

Again, in terms of ensuring aligned support (Domain 2 in Figure 11.1), if a course design includes interactive, online learning like the use of augmented reality apps as part of our ESD living laboratory initiatives, the capacity of the university's ICT support systems and infrastructure to deliver this effectively must be confirmed before the course is approved. Similarly, if the course requires staff with a particular profile, their availability to deliver the course must be confirmed before the course is approved.

This is why the various domains in Figure 11.1 are shown to overlap. A key resource in UWS courses is the provision of a learning guide for each unit of study. These self-teaching guides first identify the capabilities to be developed and then, upfront, give the assessment tasks that will be used to measure their development, along with specific explanation of how grading works, with examples. Finally, each self-teaching guide then tells the students how the various learning experiences and resources built into the unit will help them complete their assessment tasks. Peer support has been consistently found to be a key ingredient and its use to enhance retention is leveraged in each programme, especially the use of more senior students from the same equity group working with those in the first year of their studies.

Assuring graduate quality: identifying and validating the capabilities that count[1]

Programme impact (Domain 4 in Figure 11.1) is currently attracting increased focus around the world. Here, a key question is, 'How can we be sure that the capabilities given focus in a particular university degree are those which are most relevant and desirable for effective professional or disciplinary performance and constructive nation-building?'

We have found that this important aspect of L&T practice is often relatively underdeveloped and is not always jointly reviewed by those teaching the same course in different universities.[2] Box 11.1 identifies the mix of university and external reference points that can be used in Australia to assist this process:

Box 11.1 Reference points that can be used to validate learning outcomes in higher education learning programmes

An appropriate mix of the following reference points can be used:
- the discipline standards produced by the Australian Learning & Teaching Council (now the Office for Learning & Teaching);
- the UK subject benchmarks along with the outcomes of the European Tuning Project and the OECD's AHELO project;
- external professional accreditation standards (when applicable);
- data on learning outcomes in the university's annual course reports;
- the results of faculty/school reviews, especially recommendations concerning future positioning in the discipline or profession concerned;
- the stated learning outcomes for courses of the same name in other universities that are attracting high ratings on the national HE course-experience questionnaire;
- the university's graduate attributes and mission;
- the results of the studies of successful early graduates in the area concerned;
- the results of higher education employer surveys and employer round-tables on emerging opportunities for sustainability specializations in the professions;
- input from external course-advisory committees;
- national policy requirements where these exist (e.g. a requirement that all graduates should be sustainability literate and change-implementation savvy, and have critically appraised the key indicators being used to evaluate national progress and well-being).

A key issue for universities and colleges over the next decade will be to resolve whose voice will be given most (and least) weight as the learning outcomes (the capabilities and competences to be developed)

are determined for each course of study. The various voices include those of academics, the professions, the government, successful graduates, incoming students, parents, the university via its mission and Act, employers, accreditation bodies and other regional stakeholders, including those active in the area of education for sustainable development.

Assessing graduate capability: validly

Figure 11.2 identifies an empirically tested professional capability framework that can be used to validate course learning outcomes. It is important to note that the 'higher order' personal, interpersonal and cognitive capabilities identified in the top three circles have repeatedly been identified as being critical for effective early-career professional and disciplinary practice in the studies undertaken with successful early-career graduates in nine professional areas and in surveys of employers.[3] Equally, the possession of relevant generic skills and knowledge and of profession- or discipline-specific knowledge (the bottom two circles) is important. However, it has been found that simply possessing the right skills and knowledge is necessary but not sufficient for effective early-career practice and a productive contribution to society.

It is therefore important to distinguish between the terms 'capability' and 'competence', as they are often used interchangeably but incorrectly:

> Whereas being competent is about delivery of specific tasks in relatively predictable circumstances, capability is more about responsiveness, creativity, contingent thinking and growth in relatively uncertain ones. What distinguishes the most effective (performers)...is their capability – in particular their emotional intelligence...and a distinctive, contingent capacity to work with and figure out what is going on in troubling situations, to determine which of the hundreds of problems and unexpected situations they encounter each week are worth attending to and which are not, and then the ability to identify and trace out the consequences of potentially relevant ways of responding to the ones they decide need to be addressed.... While competences are often fragmented into discrete parcels or lists, capability is a much more holistic, integrating, creative, multidimensional and fluid phenomenon. Whereas most conceptions of competence concentrate on assessing demonstrated behaviours and performance, capability is more about what is going on inside the person's head. (Scott et al., 2008, p. 12)

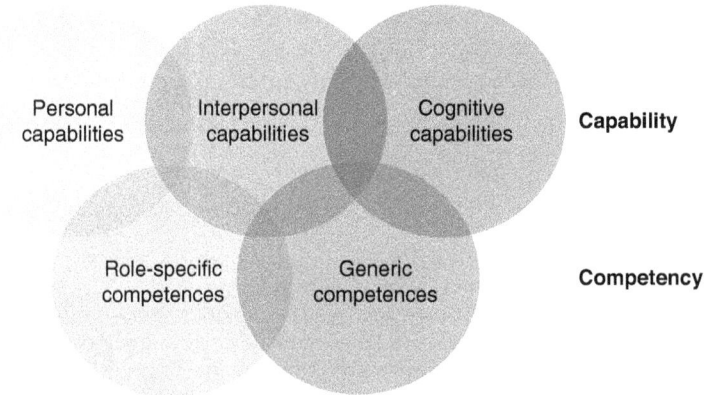

Figure 11.2 Professional capability framework

And, as Stephenson (1992, p. 1) concluded more than 20 years ago, capability depends 'much more on our confidence that we can effectively use and develop our skills in complex and changing circumstances than on our mere possession of these skills'.

Studies of successful early-career graduates in nine professions

One of the reference points identified in Box 11.1 is a set of studies of successful early-career graduates that has been undertaken over the past decade in nine professional or disciplinary areas: accounting, architecture, education, engineering, IT, journalism, law, nursing and sports management.[4] These studies focus on graduates in their first three to five years of professional practice who are identified by their clients, colleagues and employers as the most effective performers. The research invites these 'successful' graduates to identify the capabilities and competences that they believe are most important for effective performance, along with case studies of when their professional capability has been most challenged and how they managed this situation. They are also asked to look back at their university experience to identify what learning strategies, forms of assessment and experiences have, in hindsight, proven to be most (and least) beneficial.

The results are then used to 'backward map' (Elmore, 1979) to every section of the Academic Standards Framework (Figure 11.1) to help ensure that each section is demonstrably relevant to what will actually be needed specifically in the field concerned when students graduate.

The capability and competency framework depicted in Figure 11.2 comprises 38 items. Box 11.2 lists the top 12 ranked items on importance across all the successful graduate studies undertaken to date:

Box 11.2 Successful graduate studies

Top 12 capabilities by importance out of 38 (> 4.4/5, all professions) rank order, highest first
- Being able to organize work and manage time effectively (GSK)
- Wanting to produce as good a job as possible (P)
- Being able to set and justify priorities (C)
- Being able to remain calm under pressure or when things go wrong (P)
- Being willing to face and learn from errors and listen openly to feedback (P)
- Being able to identify the core issue from a mass of detail in any situation (C)
- Being able to work with senior staff without being intimidated (IP)
- Being willing to take responsibility for projects, including how they turn out (P)
- Being able to develop and contribute positively to team-based projects (IP)
- A willingness to persevere when things are not working out as anticipated (P)
- The ability to empathize and work productively with people from a wide range of backgrounds (IP)
- Being able to develop and use networks of colleagues to help solve key workplace problems (IP)

Code: P – Personal capabilities; IP – Interpersonal capabilities; C – Cognitive capabilities; GSK – Generic skills and knowledge

---

What these findings consistently show is that emotional intelligence (personal and interpersonal capabilities) and a particular diagnostic and contingent way of thinking are what count. For example, of the top 12 highest-ranked items in Box 11.2, nine concern emotional intelligence and two concern cognitive capability. Importantly, these capabilities also have much in common with the characteristics of change-capable, resilient, sustainable organizations and societies; they also have much in common with the underpinning values of the world's religions.

The findings align well with the findings from our recent international study of turnaround leadership for sustainability in higher education (Scott et al., 2013). In that study we found that the key capabilities necessary for graduates to navigate the social, cultural, economic and environmental challenges of the 21st century included being not only highly competent in their chosen profession or discipline but also sustainability literate, change-implementation savvy, inventive, capable of systems thinking, able to work across disciplines and productively with a diverse range of people to achieve a desired outcome, and having a considered position on the tacit assumptions currently driving much of the 21st-century agenda: assumptions like growth is good, consumption is happiness, technology is the answer and globalization is best.

The issue is then the extent to which such attributes are given focus in the curriculum, support systems, delivery and assessment of degree programmes.

Assessing graduate capability

We know that what is given focus in assessment in universities is generally what students most emphasize in their learning. It is, therefore, important to ensure not only that the capabilities developed are the ones that count but also that how they are assessed and graded is valid (fit for purpose). For example, it would generally be invalid to measure the development of capabilities like being able to diagnose and tackle effectively what is going on in a problematic, real-world practice situation in the area of sustainable development with complex technical, ethical and human dimensions by using a short answer or objective test or by asking students to regurgitate facts and lecture notes in a written examination.

This issue of ensuring that assessment tasks and the marking schemes used to grade them are valid is currently the subject of a national peer-moderation project involving 12 Australian universities (Krause and Scott, 2014) and in a National Senior Teaching Fellowship being undertaken by the author with the Australian Government's Office for Learning and Teaching.

What is emerging from this work is that the key capabilities in Box 11.2 *can* be measured validly and efficiently but that this requires assessment tasks and tools that are different to those typically used in many traditionally structured higher education programmes. Examples of more relevant assessment strategies include the use of:

- integrated assessment tasks and multimedia case files based on the actual real-world sustainability dilemmas faced by graduates in the first three to five years of professional or disciplinary practice;
- capstone assessment tasks involving students being assessed in real-world contexts as they seek to support workplace or community change in areas like sustainability;
- the practicum to check the capabilities that count, especially the extent to which, when students are most challenged on placement, they effectively demonstrate the highest-ranking items on emotional

244 *Geoff Scott*

and cognitive capability identified by successful early-career graduates in the profession concerned; and
- simulations, hypotheticals and trigger films as part of the assessment regime.

EfS assessment lends itself nicely to the use of such techniques.

One assessment strategy well suited to measuring key personal and interpersonal capabilities involves the use of trigger films. For example, in medicine, a large number of students in an examination hall is shown a trigger film in which a general practitioner is seen sitting in a surgery while, outside, a young mother with two happy children is waiting for her appointment. The general practitioner is shown opening an envelope that contains the results of a recent set of tests for breast cancer indicating very bad news for the young mother. The video is stopped and the students are asked to write down how they would break the news to the young mother; their answers are collected. The rest of the video in which the general practitioner breaks the news to the young mother is then shown. When this is complete, the students are asked to evaluate how well the general practitioner handled the situation and to compare what the doctor did with what they said they would do.

The *how* of turnaround: key lessons on successful change implementation in higher education

As noted earlier, the key motto for achieving L&T performance improvements at universities like UWS over recent years has been to keep in mind that 'good ideas [as discussed above] with no ideas on how to implement them are wasted ideas'. It is to this relatively underdeveloped aspect of higher education operations – how to make sure that a desired change like an new EfS programme is actually put into practice successfully and sustained in ways that benefit students – that we now turn.

Nine key change-implementation lessons have been learnt from 40 years' experience in trying to take desired L&T innovations and enhancements and make them work. Each is supported by findings from extensive research on this area across the world and can be applied directly with innovations in the area of education for sustainability (Scott et al., 2013).

It is important to note that the performance improvements identified in the first part of this chapter have taken between three to five years to implement. This time frame for implementation is typical and the implementation process unfolds in a cyclical rather than in a linear fashion.

Quality Management of Education for Sustainability 245

Key lessons on taking a desired EfS initiative and ensuring effective and consistent implementation

Consensus around the data, not around the table

A change-capable and resilient university culture focuses on what the data from the university's Tracking and Improvement System for L&T (TILT) says most needs improvement action, not, as is often the case, on consensus around the table via personal anecdote.

Set a small number of agreed priorities for action

There are too many aspects of higher education practice to be improved for there to be time to do them all: so a small number of agreed enhancement priorities must be set. One way to do this in the L&T area is to select those areas that rate very high on importance in student feedback surveys (that is, areas rated greater than or equal to 4.2/5 on a five-point Likert scale) but which attract much lower ratings on performance (less than or equal to 3.2/5).

Steered engagement

Steered engagement has emerged as being a critically important strategy to use in a university context and is a key mechanism for engaging local staff (Fullan and Scott 2009, pp. 85–88). It involves jointly setting a small number of priorities (e.g. embedding EfS more consistently into the university's programmes) and then asking each local group to work out how best to address that priority given local resources, history, circumstances, capability, students and the field of education concerned. 'Steered engagement' seeks, therefore, to give overall coherence but allows local responsiveness and variability within an agreed set of overall parameters. This combined top-down and bottom-up approach is far more productive than the 'one-way', 'top-down', 'one-size-fits-all' approach that is often used.

'Why don't we', not 'why don't you'

This is another distinguishing characteristic of a change-capable and resilient university culture. Once you hear staff saying 'why don't we', you will know that the university is on the right track. People saying 'why don't you' or 'why doesn't the university do it' transfer the responsibility for the local action that is so crucial for successful implementation (behaviour change) to someone else.

Change is learning

It is a myth that change is an event, like the launch of a new EfS plan, course or policy. Rather, it is a complex learning and unlearning process for all concerned. If nothing new is to be done – and therefore to be learnt – there is no change, simply window-dressing.

How higher education staff like to learn is how students like to learn

Motivators for staff to engage in learning (that is, to develop their capabilities consistent with the aims of a given change) are both:

- extrinsic (for example, staff knowing that the university must undergo an external quality audit and that they need to be able to show how they are actioning the changes being given focus in the audit, being rewarded for the successful implementation of a desired EfS program and so on);

and:

- intrinsic (for example, staff having a strong 'moral purpose' like wanting first-in- family students to succeed, wanting to help their nation address the challenges of climate change or achieving societal well-being, or experiencing a positive response from their students).

Like their students, staff respond to the same quality-learning experiences summarized in the RATED CLASS A checkpoints for quality-learning design identified in Domain 1 of Figure 11.1.

Again like their students, staff having to learn something new most value 'just-in-time' and 'just-for-me' solutions that specifically help them to tackle the agreed gaps in their expertise.

A key resource for staff learning is having access to what 'successful travellers' traversing the same change path in other locations have learnt works best in that specific area of practice. There is great potential to link and leverage the existing EfS networks across the world to provide this support. Such networks include AASHE in North America, COPERNICUS in Europe, ACTS in Australia and, more broadly, the International Association of Universities and the network of more than 130 higher education institutions that host the UNU Regional Centres of Expertise in ESD.

Peer group counts – the impact of their peer group on staff (and student) engagement and learning when change is in the air cannot be overestimated. This is why building a change-capable, 'why don't we', proactive culture is so important.

Knowing 'where I fit into the bigger picture' and being acknowledged for a job well done

Knowing where they fit in motivates staff to persevere with the learning and unlearning that accompany every change effort, including in the area of EfS innovation. The framework in Figure 11.1 has been used to assist staff – both academic and professional – to see that they all have an important role to play in supporting change in EfS and subsequent student success.

Learning from others

Learning from others includes targeted benchmarking with like universities. Using a common L&T tracking system like TILT helps to identify successful improvement solutions beyond one's own institution. UWS has undertaken this work with similar universities that have a similar profile within and beyond Australia. The framework in Figure 11.1 could be used to develop an integrated network of EfS innovators across the world: a network that could use an evidence-based system like the UWS TILT to identify and share good practice.

Using a government quality audit as an external lever for internal (culture) change and improvement

The very areas that are given priority focus in higher education quality audits are exactly the same as those that characterize the most change-capable and resilient colleges and universities (and, incidentally, the most effective university leaders):

- consistency and equivalence;
- a focus on positive student outcomes and not just on inputs;
- acknowledgement of what has been achieved and a concentration on what still needs to be done;
- prompt and effective action on agreed improvement areas;
- recognition that everyone has a role to play in ensuring that students have a productive university experience;
- the active use of external benchmarking and invited reviews to identify key areas for further improvement.

UWS Tracking and Improvement System for Learning & Teaching (TILT)

As noted above, a key strategy for improving L&T performance is to ensure that there is consensus around timely, valid, agreed tracking data.

The UWS TILT is similar to many systematic tracking systems now being used by higher education institutions around the world and can be used to track and improve the quality of EfS innovations.

The key characteristics of the UWS TILT system are as follows:

- Items focus on what counts, that is they focus on the key, empirically validated, quality checkpoints identified in Figure 11.1.
- In the UWS student satisfaction survey, which covers the total UWS experience (all of Figure 11.1), respondents rate each item not only on performance but also on importance. This enables the university to target its valuable resources on those areas that rate high on importance but low on performance.
- Everyone knows that there is an agreed performance standard of 3.8/5 (70% explicit satisfaction). A rating of 3.8/5 simply means that more students rate the aspect concerned with 'Agree' (4) or 'Strongly agree' (5) on the five-point Likert scale used rather than 'Neutral' (3), 'Disagree' (2) or 'Strongly disagree' (1).
- Qualitative as well as quantitative data are used to identify improvement priorities and solutions (UWS now has some 500,000 best aspect/needs improvement comments from its students, which are analysed to identify improvement hotspots using the CEQuery qualitative analysis software).
- Every course team receives an annual course-diagnostic report and a suggested action plan. Busy staff receive an integrated diagnosis undertaken for them by the university's Office of Strategy and Quality on where a triangulation of TILT data indicates that improvement is necessary. They are then invited to test the veracity of the diagnosis against an accompanying data portfolio. Improvement plans are signed off and performance on them is tracked by the Education Committee of Academic Senate.
- The next time that each unit of study is delivered, staff are asked to tell their students in the first class what actions identified from the last cohort of students are being taken to improve the unit. It has been found that, when this is done, students are more likely to respond to the next round of surveys because they can see their feedback being acted on. It also helps to ensure that staff act on the areas of improvement agreed for their unit of study.
- The system is ICT-enabled. This gives a fast, timely turnaround of results and enables the university to identify units of study or programmes quickly where a particular activity or support system is attracting very high performance ratings. These exemplars are used to

identify practical improvement ideas for staff in units of study where the performance ratings for the same L&T aspect are much lower. This concept of 'benchmarking for improvement' is much valued.

Change does not just happen but must be led, and deftly

In the *Learning Leaders in Times of Change* study of some 500 experienced learning leaders in higher education (Scott et al., 2008), it became clear that changes like those outlined above do not just happen but that they must be led – and deftly. The key lessons from that research match the findings from the broader set of studies of effective change leadership in universities and colleges identified in *Turnaround Leadership for Higher Education* (Fullan and Scott, 2009) and, more recently, in the international study of *Turnaround Leadership for Sustainability in Higher Education* (Scott et al., 2013).

Importantly, the lessons below apply to everyone who has a leadership position – both local and central. The key lessons for effective turnaround leadership for sustainability in higher education are as follows:

Listen, link, then lead – always in that order

Always *listen* with a framework. This includes providing a rationale for introducing a change like EfS and options that have proven successful in addressing it elsewhere. Listening with a framework involves asking those who are being invited to implement the EfS initiative what they think, on the evidence given, would be a most relevant, desirable and feasible (achievable) way to address it. *Link* involves bringing together the most common responses generated during the listening phase and *lead* involves helping those who are to implement the change learn how to do so. This 'listen, link, then lead' approach is consistent with the notion of 'steered engagement' discussed earlier.

Model, teach and learn

The most effective higher education leaders model the attributes of a change-capable culture to their staff – especially how to behave when things go wrong or the unexpected happens. Like successful early-career graduates, effective higher education leaders have high levels of emotional intelligence and display a contingent, diagnostic way of thinking. In modelling these capabilities, they show others how to manage change and uncertainty in a constructive, collaborative, focused, adroit and practical fashion.

Equally, the most effective leaders constantly try to improve their own practice and capabilities (that is, they continue to learn). They do this by using the professional capability framework in Figure 11.2 and the results from studies of successful HE leaders in roles like theirs, along with carefully developed networks of fellow practitioners (Scott et al., 2008; Scott et al., 2013).

A change-capable culture is built by change-capable leaders

If both local and central L&T leaders intentionally model the personal, interpersonal and cognitive capabilities identified as distinguishing effective L&T and EfS leaders, other staff pick this up as 'the way we do things around here'. It is in this way that change-capable cultures are built: not by talking but by showing people what works. This approach is based on the premise mentioned earlier that 'we are more likely to act our way into new ways of thinking than think our way into new ways of acting' (Spohn, 2003).

Everyone is a leader in her or his own area of expertise and responsibility

It is a myth that leadership resides only with the most senior people in a university. In fact, a change like building EfS into the curriculum will not be put into practice consistently and well if local leaders, like heads of school, local managers and heads of programme, do not actively engage with it and help their staff learn how to do it. This has been confirmed in our recent study of turnaround leadership for higher education in three continents (Scott et al., 2013).

Our capability is most tested when things go wrong or the unexpected happens

We learn most and our capabilities as a leader are most tested when things go wrong or the unexpected happens. The specific findings on the capabilities that count for every L&T leadership role (Scott et al., 2008) have been used for targeted, role-specific support programmes. A specific profile has been developed for EfS change leaders in both central and local roles in the turnaround leadership for EfS study (Scott et al., 2013)

Key change myths

As you seek to develop a change-capable and resilient culture and to shape and implement a systematic approach to embedding EfS in your higher education institution, it may be helpful to keep the following change myths in mind:

- *The knight-on-the-white-charger myth:* this refers to the myth that all one needs is one charismatic leader and all will be well.

- *The brute-sanity myth:* as both George Bernard Shaw and Albert Einstein are said to have observed: reformers have the misplaced idea that change is achieved by brute logic.
- *The restructure myth:* this refers to the myth that restructuring will automatically improve quality and ensure successful change implementation.
- *The change-event myth:* this refers to the myth that passing a policy, launching a plan or signing off on a new course means that it will automatically be effectively and consistently adopted and implemented. Rather than being an event, however, change is a complex learning and unlearning process for all concerned.
- *The 'why don't you' myth:* this myth transfers responsibility for action away from the frontline to someone else. In change-capable universities, this is replaced by consistent talk of 'why don't we'.

Summary

The following five aphorisms summarize the characteristics of the change-capable culture now necessary for universities to navigate the 21st-century challenges outlined at the outset of this paper successfully and, in doing this, give more consistent focus to social, cultural, economic and environmental sustainability in their core activities:

- 'We rise to great heights by a winding staircase' (Francis Bacon).
- 'People are more likely to act their way into new ways of thinking than think their way into new ways of acting' (Spohn, 2003).
- Ready (see a need for change), fire (try out and refine a solution under controlled conditions), then aim (identify what works) – rather than ready, aim, aim, aim, aim…have another meeting.
- Listen, link, then lead – always in that order.
- Effective leaders model, teach and learn.

Conclusion

Good ideas with no ideas on how to implement them are wasted ideas.

This chapter has argued that colleges and universities across the world are collectively faced by a linked set of change forces that require both new approaches to learning and teaching as well as a more systematic way to develop, monitor and enhance their current learning programmes. It has been argued that the institutions that will flourish and best serve their students and nations in the new context of the 21st century will be very clear on what engages and retains students in productive learning and, most importantly, will have the capacity and know-how to put desired changes and quality improvements into practice – both consistently and effectively.

That is, it will be those institutions that understand not only the what of effective learning and teaching and education for sustainability but the how of making change work in ways that will gain and retain students and produce quality graduates to the benefit of both the individuals concerned and their nations.

The strategies discussed in this chapter can be used not only to prove but improve the quality of what is being designed and delivered in higher education, including in the area of education for sustainability. In this regard an important initiative which is currently under way in the UK under the leadership of the University of Gloucestershire and supported by the UK Higher Education Funding Council and the UK Quality Assurance Agency is the project focused on 'Leading Curriculum Change for Sustainability: Strategic Approaches to Quality Enhancement'.

Universities now have the crucial role in producing future change leaders who can effectively tackle the fundamental challenges of our time: how best to build social, cultural, economic and environmental sustainability. The L&T quality framework, the TILT system and research on effective change management and turnaround leadership for HE discussed in this chapter will hopefully give some focus to this critical work and ensure that what is attempted is evidence-based, builds on existing good practice and actually has a positive impact.

Notes

1. A more detailed paper and videotape on the issue of researching graduate capability were presented at the South African Cape Higher Education Consortium Forum in 2011. A video on the issues discussed in this chapter was also produced for the HEFCE/QAA conference on developing a quality management framework for UK education for sustainability programmes in UK universities.
2. For a national study that explores this area see Krause and Scott (2014).
3. For examples of these studies, see Scott and Wilson (2002); Scott and Yates (2002); Scott and Wilson 2002, Wells et al. (2009); Scott et al. (2010) and the *survey section* of the UWS TILT website.
4. See endnote 3.

References

Dewey J. (1933) *How We Think: A Restatement of the Relation of Reflective Thinking to the Educative Process* (Boston, MA: Heath).
Elmore R. F. (1979) 'Backward Mapping: Implementation Research and Policy Decisions'. *Political Science Quarterly*, 96, 601–616.
Fullan M., Scott G. (2009) *Turnaround Leadership for Higher Education* (San Francisco: Jossey Bass).
Krause K.-L., Scott G. (2014) *Assuring Learning and Teaching Standards through Inter-institutional and Peer Review* (Sydney: Australian Government, Office of Learning and Teaching).
Scott G. (1999) *Change Matters: Making a Difference in Education and Training* (Sydney and London: Allen & Unwin).
Scott G. (2006) 'Accessing the Student Voice: Using CEQuery', Department of Education, Employment and Workplace Relations, Canberra, http://www.uws.edu.au/__data/assets/pdf_file/0010/63955/HEIPCEQueryFinal_v2_1st_Feb_06.pdf, date accessed 31 May 2012.
Scott G. (2008) 'University Student Engagement and Satisfaction', report commissioned for the *Bradley Review of Australian Higher Education*, Australian Government, Canberra, http://logincms.uws.edu.au/__data/assets/pdf_file/0008/78668/Research_HE_review_0908_Scott.pdf, date accessed 7 August 2014.
Scott, G. (2013) 'Improving Learning and Teaching Quality in Higher Education'. *South African Journal of Higher Education*, 27 (2) 275–294.
Scott G., Chang E., Grebennikov L. (2010) 'Using Successful Graduates to Improve the Quality of Undergraduate Nursing Programs'. *Journal of Teaching and Learning for Graduate Employability*, 1 (1) 26–44.
Scott G., Coates H., Anderson M. (2008) *Learning Leaders in Times of Change* (Sydney: Australian Learning and Teaching Council).
Scott G., Hawke I. (2003) 'Using an External Quality Audit as a Lever for Institutional Change'. *Assessment & Evaluation in Higher Educations*, 22 (3) 323–332.
Scott G., Tilbury D., Sharp L., Deane E. (2013) *Turnaround Leadership for Sustainability in Higher Education: An International Study* (Sydney and Canberra: Australian Government, Office of Learning & Teaching).
Scott G., Wilson D. (2002) 'Tracking and Profiling Successful IT Graduates: An Exploratory Study', ACIS Proceedings, http://aisel.aisnet.org/acis2002/92, date accessed 31 May 2012.
Scott G., Yates W. (2002) 'Using Successful Graduates to Improve the Quality of Undergraduate Engineering Programs'. *European Journal of Engineering Education*, 27 (4) 363–378.
Spohn W. (2003) 'Reasoning from Practice'. Carnegie *A Life of the Mind for Practice* seminar, Stanford, CA, Carnegie Foundation for the Advancement of Teaching.
Stephenson J. (1992) 'Capability and Quality in Higher Education'. In J. Stephenson, S. Weil (eds) *Quality in Learning* (London: Kogan Page).
Sullivan W., Rosin M. (2008) *A New Agenda for Higher Education: Shaping a Life of the Mind for Practice* (Thousand Oaks, CA: Corwin).

Tough A. (1977) *The Adults Learning Projects*, 2nd edn (Toronto: OISE Press).
Vescio J., Scott G. (2005) *Studies of Successful Graduates* (Sydney: UTS).
Wells P., Gerbic P., Kranenburg I., Bygrave J. (2009) 'Professional Skills and Capabilities of Accounting Graduates: The New Zealand Expectation Gap?' *Accounting Education*, 18 (4) 403–420.

12
Implementing Education for Sustainable Development in Higher Education: Case Study of Albukhary International University, Malaysia

Salfarina Abdul Gapor, Abd Malik Abd Aziz,
Dzulkifli Abdul Razak and Zainal Abidin Sanusi

Introduction

Monitoring and evaluation (M&E) are an essential component of the governance for any institute of higher education, especially in a multifaceted area such as sustainability. The measuring criteria and the tools needed to conduct the assessment process are entirely dependent on the purpose for which the assessment is being carried out, and the objectives and scope of the assessment should be well defined before the assessment takes place. In the case of sustainability assessment, the aim should be creating a balance between the key elements of sustainability, namely environment, society and economy. The needs and concerns of each institute of higher education may vary, however the above three elements can be considered as the core components of any sustainability assessment globally.

Conditions such as financial limitations are factors in deciding how deep or detailed the assessment should be. The expertise of those conducting the assessment should also be a matter of high consideration; on the other hand, there should also be a rough expectation of how cooperative the assessed entity will be with the assessors. Since the assessment process requires massive data collection, bilateral cooperation is of vital importance. In addition, the parameters for each assessment project should be directly relevant to the condition, context and setting of the institute of higher education concerned.

In terms of the quality indicators and criteria for the assessment, various parameters can be considered, such as: research excellence and/or influence, student choices, eventual success and/or demographics and others based on the institution's mission and strategic objectives.

Assessing the quality of the institutions of higher education based on a set of sustainability criteria is gaining increasing momentum worldwide. There are several guidelines and methodologies developed by different institutions in this regard. However, 'Although these documents contain important guidelines for education, none of them offers concrete prescriptions on an operational level for what higher education should do exactly in order to contribute maximally to sustainable development' (Shriberg, 2002b).

In order to overcome these gaps, a number of projects from around the world have proposed certain initiatives geared toward reforming the current assessment systems of HEIs, especially in terms of sustainability principles, namely the Alternative University Appraisal (AUA), the Sustainability Tracking, Assessment and Rating System (STARS) and Sustainable Endowment Institute (SEI). The development of these tracking and assessment systems represents a positive step towards mainstreaming sustainability in the core activities of higher education institutes, provided the quality of the sustainability elements and criteria in the systems reflects the true meaning of sustainability and the feasibility of local implementation. The successful delivery and success of such systems depend heavily on political will and active support of global academia.

Alternative University Appraisal (AUA)

The AUA model is one of a number of projects that has emerged from a network of universities called ProSPER.Net: the Promotion of Sustainability in Postgraduate Education and Research which was initiated in 2009 to appraise universities via an alternative set of perspectives linked to sustainability. AUA seeks to:

> facilitate and encourage institutions of higher education to engage in education and research for sustainable development and to raise the quality and impact of these activities by providing benchmarking tools that support diversity of mission, as well as a framework for sharing good practices and supporting dialogue and self-reflection. (Senaha, 2010)

A fundamental goal of the AUA is to focus less on the ranking of universities and place greater emphasis on the rating of universities. The aim of the project is not to propose an appraisal system for a small subset of universities that reject the mainstream ranking systems and wish to choose an alternative path, but it is to advocate the empowerment of an institution of higher education to decide a relevant, feasible and focused development strategy for itself. The process involves self-reflection between partnering institutions using three procedures – self-awareness questions (SAQ), a dialogue, and benchmark indicator questions (BIQ) – which enable HEI to assess their individual ESD activities and encourage self-awareness of their own strengths/weaknesses in the field of ESD. Both SAQ and BIQ focus on assessing four dimensions: governance, education, research and outreach. For the SAQ, the data collected and assessed are mainly of a qualitative and subjective data covering specific ESD activities, gender, and so on. The BIQ data are more quantitative and objective, and seek to determine the overall sustainability maturity of the university. Those involved in the assessment include faculty, staff, students and other relevant stakeholders. Both BIQ and SAQ provide self-reflection opportunities regarding sustainability activities and complement each other in order to ensure that both obvious and latent activities, of both specific and overall scope, are taken into consideration.

AUA provides two benefits. It enhances the value and attractiveness of universities engaging in ESD and it creates a supportive learning community to improve their practices. Given the holistic and flexible criteria of the AUA model, universities can aspire to higher ratings according to both conventional and education for ESD measures. In addition to this, the AUA system recognizes the good practices of participating universities that consciously espouse the principles of ESD and also aims to shape the ways in which universities operate for a more sustainable future by recognizing diversity, innovation and successful change towards sustainable development. Hence, AUA functions along the vein of other alternative appraisal systems such as that provided by AASHE, the ICHE Observatory Project and the University Rating System for ASEAN/Southeast Asia which is currently being developed.

The Sustainability Tracking, Assessment & Rating System (STARS)

The Association for the Advancement of Sustainability in Higher Education (AASHE) introduced the Sustainability Tracking, Assessment & Rating System (STARS) in 2006 to promote a sustainability agenda in higher education institutions at all levels in North America, with a pilot rating exercise also available for universities outside this region. As discussed in the earlier, more detailed chapter on the system (Chapter 8), the main strength of STARS is its ability to compare performance of institutions over time and across institutions using standardized assessments. Being able to claim a high STARS rating helps an institution reap the marketing and recruitment benefits of being a sustainability leader. It also promotes innovation in sustainability and facilitates information sharing on sustainability practices and performance.

Institutions earn points in three main categories: Education & Research; Operations; and Planning, Administration & Engagement. Each of these categories includes subcategories such as Purchasing, Curriculum, Energy, and Human Resources. There is also an Innovation category to recognize pioneering practices that are not covered by other STARS credits. Like AUA, its emphasis is self-assessment and on rating rather than ranking. The office in charge of sustainability in each participating university collects the sustainability data and feeds it into the online STARS reporting tool. The final rating, based on the point system associated with the indicators, can be used to compare that university against other universities, or to compare the university against its own progress over time. Institutions that wish to participate in STARS but do not want to publish their scores may participate as a STARS Reporter. Institutions that wish to be scored may earn one of four other levels of STARS ratings: Bronze, Silver, Gold and Platinum. Once an institution has earned a credit, it is expected that the institution will maintain the status that made them eligible for the credit during the duration of the STARS rating. While fluctuations in some performance areas are to be expected, an institution would not qualify for a credit if it ended the practice or policy upon earning the credit and resumed the practice or policy when it was time to re-submit information.

Sustainable Endowment Institute (SEI)

The Sustainable Endowment Institute (SEI) evaluates universities in North America, based on their sustainability-related activities through the College Sustainability Report Card (SEI, 2011). SEI is different from AUA and STARS because of its lack of emphasis on research and teaching and its focus on sustainability practices in areas like administration, climate change and energy, food and recycling, green building, student involvement, transportation, endowment transparency, investment priorities and shareholder engagement. Each of these categories is divided into subcategories with specific criteria. For example, in terms of climate change and energy, the criteria are: a greenhouse gas emissions inventory, commitment to greenhouse gas emissions reduction, realized greenhouse gas emissions reduction, energy efficiency and conservation, renewable energy generation, renewable energy purchase, and on-site combustion.

In terms of the assessment system, SEI follows the traditional approach of selection, survey composition, data collection and verification, assessment and recognition. Universities are awarded points based on their levels of activity within each indicator, with extra credit points awarded for highly innovative efforts. When appropriate, school size and geographic setting are taken into account. Within each of the nine categories, universities that receive at least 70 per cent of the available credit earn an A, 50 per cent earn a B, 30 per cent earn a C, and 10 per cent earn a D. The nine main categories are weighted equally to calculate the GPA on a 4.0 scale, which is then converted into an overall letter grade.

Strengths and weaknesses of AUA, STARS and SEI

A closer look at the above-mentioned systems of sustainability assessment in higher education identify the strengths and weaknesses of each system, mainly in the dimensions of data collection and measuring tools (see Table 12.1). For AUA, the strength is in the thoroughness of the SD activities' evaluation scope which covers both the macro (general maturity of SD 'existence' as a niche for the HEI) and micro levels (specific SD activities that cover research, teaching, community engagement and operation). In terms of data collection, the AUA assessment is sophisticated in that it combines both quantitative and qualitative data which has the advantage of helping to optimize validity and reliability. The AUA assessment methods also use two approaches to data collection, survey and dialogue, which help provide a more accurate representation of planned and implemented SD activities. In terms of assuring the overall standard of the system, AUA has been reviewed extensively by the participating ProSPER.Net HEI members through feedback and inputs from several meetings, thus making the system more relevant and inclusive for HEIs globally. The transparency of the AUA rating system also provides high visibility which, in turn, enables the sharing of ideas and strategies of SD activities.

However, there are a number of quality assurance challenges to be addressed. These include the time it takes to provide the open-ended qualitative data on the AUA survey form and the difficulties involved in ensuring the reliability and validity of its analysis. Another challenge

Table 12.1 Strengths and weaknesses of AUA, STARS and SEI

	Strengths	Weaknesses
AUA	Data collection: sophisticated method that combines 2 types of data (quantitative and qualitative) using 2 approaches to data collection: survey and dialogue; increased validity and reliability of data; more accurate; reflects the SD activities planned and implemented. Scope of evaluation: evaluation is more thorough and realistic, not only evaluating the activities but also maturity of SD implementation. Reviewed extensively: the assessment methods have received feedbacks and inputs from many HEI, thus relevant and generic. Visibility and sharing: ability to share ideas and strategies of SD activities.	Data analysis: complicated to analyse qualitative data; time consuming for stakeholders (HEI) and the secretariat; open to misinterpretation if ambiguous information is fed and not carefully verified; danger of being impractical especially for annual assessment. Data assessment process: needs more human resource and time to complete the data collection and analysis. Depth of evaluation: just assesses the existence of SD activities and not the impact of the activities.
STARS	Data analysis: activities are evaluated based on monetary values, hence easy to calculate. Innovation incentive: provides incentive to improve sustainability activities under Innovation category. Visibility and Sharing: ability to share ideas and strategies of SD activities,	Mainly focuses on US and Canada HEIs. Assessment period: every 3 years, the period is quite long, might not take into consideration new issues related to SD. Not free: need to pay registration fee to participate.
SEI	Innovation incentive: provides incentive to improve sustainability activities under Innovation category. Data collection: sophisticated method that combines 2 types of data (quantitative and qualitative) to ensure high reliability and validity of information regarding SD activities. Data collection: puts high emphasis on the accuracy of data collection to avoid biasness. Data instrument: uses different grading scales (e.g. binary, incremental) in the assessment system, varied and more sensitive measuring tools. Data grading techniques: clear indication of aggregate grading system from alphabet to percentage values. Assessment period: yearly assessment. Visibility and sharing: ability to share ideas and strategies of SD activities.	Not inclusive: only for USA and Canada. Scope of evaluation: focuses only on SD activities in campus without any outreach programmes to the community.

involves having available the increased levels of human resources, time and money necessary to provide the massive amount of data requested. Finally, although the AUA assessment tools capture SD activities, the data focus more on their existence than on their impact.

SEI has a well-developed data gathering and grading system but does not cover the core activities of a university, only the operational ones. Also it is primarily North American in focus.

For STARS, the main strength is in the data analysis and calculation method. This is because the SD activities are evaluated based on monetary values, hence it is clear, simple and easy to calculate. The STARS assessment system also provides innovation incentive for outstanding SD activities. STARS also facilitates the sharing of ideas and strategies for SD activities through their annual reports which are readily available online.

The weakness of STARS is that it is not inclusive as the assessment system is only open, at least until recently, for the Association for Advancement of Sustainability in Higher Education (AASHE) members in the United States and Canada. This is despite a new initiative to run a pilot project for other HEIs outside the United States and Canada. Furthermore, as the assessment period is for every three years, there may be a danger that new issues pertaining to SD will not be taken into account quickly enough. It is also not free, and it is cheaper for AASHE members than for non-members.

All three rating systems are developed in order to assess SD activities at tertiary level. The rating system is based on an overall picture of what aspects of SD each HEI will implement in its core and/or support activities. Hence, there is a chance that the assessment system might not take into account the specific context and needs of a particular HEI. With this in mind, the next section will discuss a sustainability assessment method that is designed for a particular HEI to meet its specific mission and vision. We will see how elements of the systems reviewed above have been adapted to suit a particular ESD innovation and the unique operating context of Malaysia's Albukhary International University.

Building a platform for sustainability assessment in higher education: the Sustainable Livelihood Approaches (SLA) used at AIU

Since its establishment in 2010 and until 2013, Albukhary University has been operating on the principle of being a 'humaniversity', a HEI where social, cultural, economic and environmental sustainability are the key focus. The AIU mission and vision is to graduate students from marginalized groups as agents of change. Alleviating poverty is one of the major goals embedded in the university's policy, and being socio-economically underprivileged is a key criterion for student selection. The vision of the university is to forge a more equitable and tolerant world through sustainable social welfare, education and cultural initiatives that bridge the gap between the rich and the poor. The university focuses on providing a learning environment that not only creates a theoretical background but also involves a planned community outreach programme that students need to undergo from their first until their final year of study. These first-hand learning experiences provide an experiential platform so that students can confront directly, learn to cope with, negotiate and successfully adapt to real life challenges.

The Sustainable Livelihood Approach (SLA) uses a partnership strategy between the community, networks of third parties and academic institutions. The programme builds knowledge, understanding and capacity from a multi-disciplinary, cross-sectoral perspective and provides training for the community to develop intellectual, cultural, social and environmental elements which directly benefit both their members and students. The SLA concept was introduced by Chambers and Conway to build the capabilities of communities in generating and sustaining their means of living, in areas including food, income and assets and the natural resources that their livelihoods depend on, as well as to build the ability of the community to cope with and recover from stress and shocks, and to provide for future generations (Chambers and Conway, 1991). Figure 12.1 gives an overall picture of how the SLA approach works.

SLA uses a combination of five approaches; Asset Building and Community Development (ABCD); Community Driven Development (CDD) approach; Rights Based Approach (RBA); Gestion de Terroir (GT) and Social Safety Net (SSN). It is an approach to sustainable development that plans and implements development projects based on the needs and strengths of the community from within, through capacity building and linking the community with the relevant agencies and stakeholders. The approach also puts a strong emphasis on quality monitoring and evaluation which is in contrast to the conventional developmental approach adopted by governmental agencies which tends to be very 'top-down' and inputs focused but lacks an ongoing system for monitoring, evaluation and continuous quality improvement.

Apart from implementing the programme, assessing the process and outputs of the programme is crucial. This involves verifying the relevance and impact of the programme on those intended to benefit from

Implementing Education for Sustainable Development 263

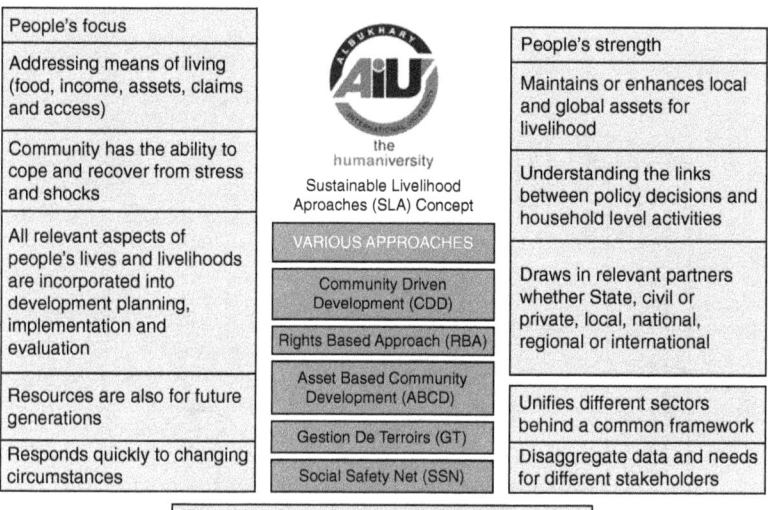

Figure 12.1 Sustainable Livelihood Approaches

it, seeing demonstrable improvements in the university's pedagogy, clear benefit for other stakeholders and most importantly a positive impact on students' capabilities, competences and achievement in assessment. The quality assessment to follow concentrates only on the SLA initiative and does not include the sustainability assessment of the whole university.

The SLA approach combines teaching, research and a systematic outreach programmes. Teaching is necessary to include students directly in the programme, as part of the curriculum and to equip them with theoretical frameworks and research methods for the outreach engagement activities. The programme also involves research, both fundamental and action-oriented. The fundamental research is undertaken at the university, while the action-research is only undertaken during the community immersion process, the implementation of action plans and monitoring and the evaluation of the projects. The key performance indicators are categorized under the three pillars of sustainable development – socio-economic, socio-cultural and environment – as presented in the assessment system known as Sustainability Rating Assessment System (SRAS) (see Figure 12.2).

Figure 12.3 shows how SLA is assessed and quality assured using the Sustainability Rating Assessment System (SRAS). The indicators are rated

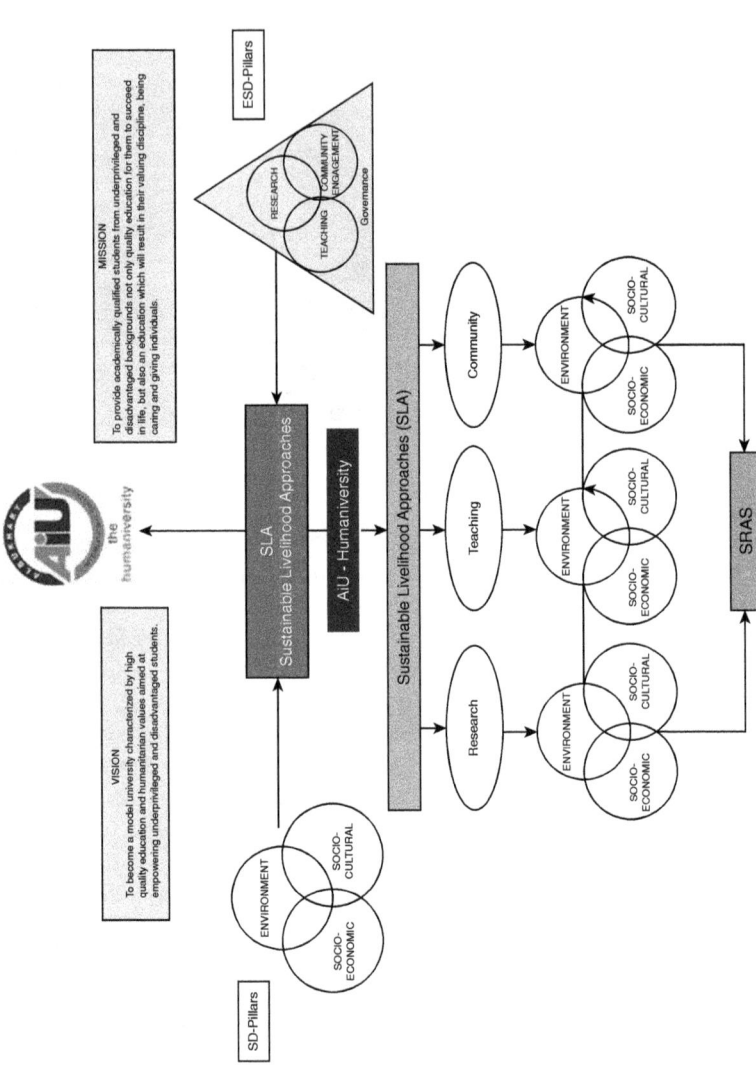

Figure 12.2 AIU/SLA framework for sustainability assessment

Implementing Education for Sustainable Development 265

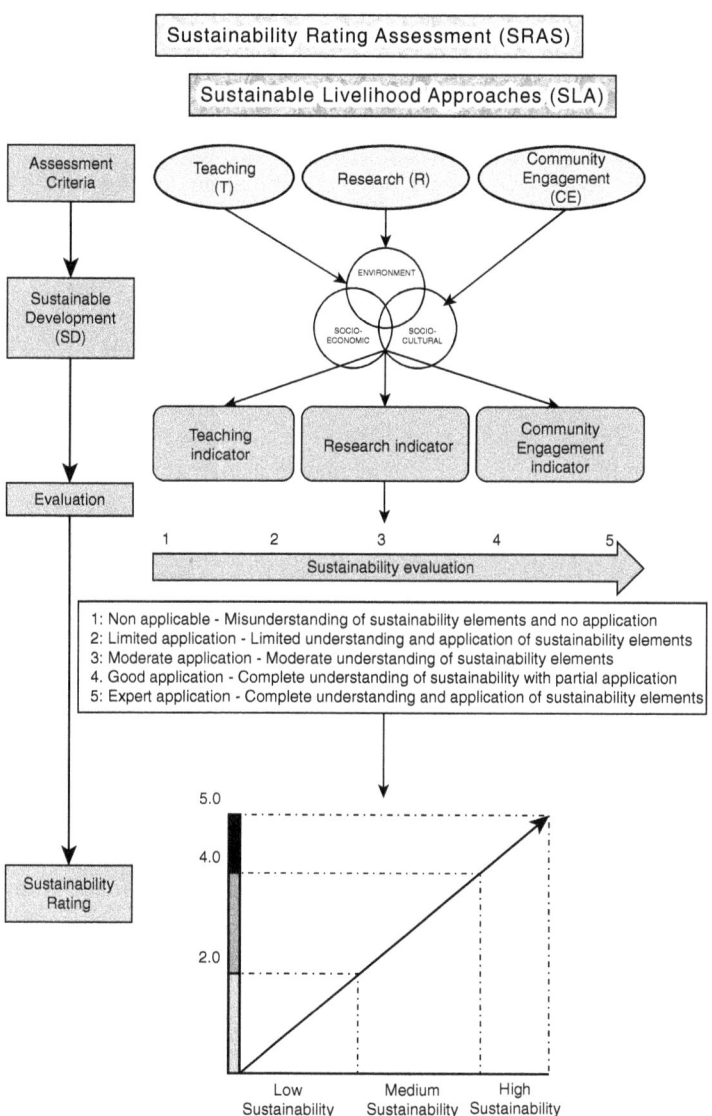

Figure 12.3 Sustainability Rating Assessment System (SRAS) for AIU/SLA programme

using a Lickert scale from 1 to 5 for every category and criterion under each of the priority settings (research, teaching and community engagement) as indicated in the key performance indicators in Tables 12.2, 12.3 and 12.4, and elaborated further in the case study. '1' is for 'non-applicable' which means that there is a misunderstanding and no application of sustainability elements. '2' is 'limited application' and understanding of sustainability elements. Both '1' and '2' reflect low sustainability elements existing in teaching, research or community engagement under the SLA programme. '3' indicates a 'moderate application' and understanding of sustainability and '4' is 'good application', meaning that there is a complete understanding of sustainability but only partial application. Both '3' and '4' reflect medium levels of focus on sustainability. '5' is the highest rating and entails 'expert application', indicating complete and full implementation of the desired sustainability elements in the targeted priority areas. An aggregate value based on the average total of all the criteria will give the total sustainability rating, from low (with scores in the range of 1 to 1.9), medium (with scores in the range of 2.0 to 3.9) to high (with scores in the range of 4.0 to 5.0).

Case study of SRAS application: the indigenous people of Kensiu, Baling, Kedah, Malaysia

The application of the assessment system is demonstrated in one of the projects with the indigenous people, the Kensiu who are part of the Negrito tribes in Malaysia. The result is based on the outcomes from the pilot project which was conducted by 11 students and a project leader. There are only 277 Kensiu left and all of the households are poor. The fact that they have been uprooted from their natural dwellings in the forest because of an insurgency which forced them to move near the small town of Siong involved them in having to interact with the mainstream ethnic groups – the Malay and Chinese – and this has posed considerable challenges for them. The AIU students undertook a 'community immersion' in order to understand the needs and problems of the Kensiu and from this identified a selection of strategies for the action. The action plans include promoting permanent integrated farming to reduce Kensiu dependency on the nearby forests for food, with the hope that this will increase food security and create an avenue for income generation. Socio-cultural projects include an education programme (aimed at helping the Kensiu both to adapt to the secular education system and to revitalize indigenous knowledge) and ethnic integration between the Kensiu and the mainstream Malay and Chinese.

Long-term programmes include creating a bamboo village concept for tourism activities and capacity building to include the Kensiu in the ecotourism project that uses the nearby forest as a main attraction.

Table 12.2 shows the four main learning and teaching quality indicators being used in the project: student quality, recognition, student involvement and visibility. For student quality, there are five output indicators which are assessed using quantitative indicators (numbers, levels, existence, grades, and so on). For recognition, the output indicator is approved and recognized by the Malaysian Quality Accreditation (MQA) agency. Student involvement is assessed through recognition and participation and valued in terms of number and level. Visibility as an output indicator is valued in terms of the frequency with which the project is reported in the media. NA stands for non-available or incomplete, since the project is still in its first year of implementation. The next columns indicate the SD pillars that are relevant to the output indicators: for example, for output indicators 'awards', the pillar includes all the three dimensions – environment, economic and social – since the project focuses and has impacts on the Kensiu under the three SD pillars. Examples of complete indicators are under 'student involvement', for example under output indicators 'number of recognition', for the Kensiu project, they have entered a competition under SIFE/Enactus (only one competition, so accumulating one point) and won the special award (five points for special award) Rookie title (best new entry/project). Total sustainability action is calculated by aggregating all the scores in the respective output indicators result. The value, for example in the case of teaching, is 25 which is divided by the total number of output indicators (13), giving the result of 1.9, which in the SRAS scale indicates low sustainability (see Figure 12.4). The result is low because indicators like 'student quality' are still incomplete and projects are still ongoing. Please refer to Table 12.2 for the more detailed calculations.

Table 12.3 identifies research indicators with three main dimensions – publication, networking/research and conferences/seminars. For publication, the output indicators are assessed based on the number and impacts of reports or journals. For networking/research, the output indicators are assessed based on number and level of networking and monetary values, while the measurement for conferences/seminars is based on number and level (national or international). The value of the research indicators is 45, which when divided by the total number of output indicators (18) gives the result of 2.5, which in the SRAS scale indicates medium sustainability (see Figure 12.4). The research component is medium because most of the fundamental research has been

Table 12.2 Teaching indicators for sustainability assessment

Priority setting	No	Indicator	Output indicator	NA	SD Env	SD Eco	SD Soc	Description	Sustainable scale 1	2	3	4	5	Total of sustainability rating (SR)-priority	SR of priority
Teaching	1	Student quality	Number of students graduated with transferable skills	√				1: 5%, 2: 10%, 3: 30%, 4: 60%, 5: 80% and above						25	1.9
	2		Quality of students graduated (CGPA, 3.00 and above)	√				1: 5%, 2: 10%, 3: 30%, 4: 60%, 5: 80% and above							
	3		Employment statistics of graduates (to include self-employed)	√				1: 5%, 2: 10%, 3: 30%, 4: 60%, 5: 80% and above							
	4		Number of student community initiatives during study period		√	√	√	1: 1–5, 2: 6–10, 3: 11–15, 4: 16–20, 5: 20 and above		2					
	5		Number of student community initiatives after graduation		√	√	√	1: 1–5, 2: 6–10, 3: 11–15, 4: 16–20, 5: 20 and above							
	6	recognition	SLA curriculum development approved and recognized by MQA				√	MQA auditing in 5 years							
	7	Student involvement	Number of recognitions (awards, competitions)		√	√	√	1: 1, 2: 2, 3: 3, 4: 5, 5: more than 7	1						

8	Quality of recognition (national/international level)	√	1: Participant, 2: 4–5th place, 3: 3rd place, 4: 2nd place, 5: champion/special award	5
9	Number of exhibitions by students	√	1: 1, 2: 2, 3: 3, 4: 5, 5: more than 7	2
10	Quality of exhibitions by students	√	1: participant, 2: 4–5th place, 3: 3rd place, 4: 2nd place, 5: champion/special award	5
11	Number of reports by students	√	1: 1, 2: 2, 3: 3, 4: 5, 5: more than 7	4
12	Quality of reports by students	√	1: Low, 2: Intermediate, 3: Medium, 4: Good, 5: Excellent	3
13	Visibility	Number of reports in the media (media, TV and internet)	√	1: 1, 2: 2, 3: 3, 4: 5, 5: more than 7	3

Table 12.3 Research indicators for sustainability assessment

Priority setting	NO	Indicator	Output indicator	NA	SD Env	SD Eco	SD Soc	Description	Sustainable scale 1	2	3	4	5	Total of sustainability rating (SR) – priority	SR of priority
Research	14	Publication	Number of research		√	√	√	1: 1, 2: 2, 3: 3, 4: 5, 5: more than 7				4		45	2.5
	15		Number of reports				√	1: 1, 2: 2, 3: 3, 4: 5, 5: more than 7				4			
	16		Number of journals	√				1: 1, 2: 2, 3: 3, 4: 5, 5: more than 7							
	17		Quality of reports				√	1: low, 2: intermediate, 3: medium, 4: good, 5: excellent			3				
	18		Quality of journals (ISI, Scorpus, citation index, etc.)	√				1: 1, 2: 2, 3: 3, 4: 5, 5: more than 7							
	19	Networking/ Research	Number of research networking		√	√	√	1: 1, 2: 2, 3: 3, 4: 5, 5: more than 7					5		
	20		Number of CSR		√	√	√	1: 1, 2: 2, 3: 3, 4: 5, 5: more than 7			3				
	21		Amount of CSR (based on Ringgit Malaysia)				√	1: RM 1–5000, 2: RM 5001–10000, 3: RM 10001–20000, 4: RM 20001–30000, 5: More than RM 30001					5		
	22		Number of grants		√	√	√	1: 1, 2: 2, 3: 3, 4: 5, 5: more than 7	1						

23	Amount of grant (based on Ringgit Malaysia)		√	√	1: RM 1–5000, 2: RM 5001–10000, 3: RM 10001–20000, 4: RM 20001–30000, 5: Lebih dari RM 30001	3
24	Level of research grant (national/ international)	√		√	1: National (RM 1–10K), 2: National (RM 11K–20K) & International (below 20K), 3: National (RM 21K–30K & above) & International (RM 31K–50K), 4: National (RM 31K–40K) & International (RM 51K–70K), 5: National (more than RM 41K) & International (more than RM 71K)	2
25	Conference/ Seminar/ Workshop		√	√	1: 1, 2: 2, 3: 3, 4: 5, 5: more than 7	2
26	Number of conferences, seminars and workshops		√	√	1: 1–50, 2: 51–100, 3: 101–150, 4: 151–200, 5: more than 201	1
27	Quality of conferences, seminars and workshops: Number of participants					
	Quality of conferences, seminars and workshops: Number of proceedings	√			1: 1–20, 2: 21–40, 3: 41–60, 4: 61–80, 5: more than 81	

continued

Table 12.3 Continued

Priority setting	NO	Indicator	Output indicator	NA	Env	Eco	Soc	Description	\multicolumn{5}{c}{Sustainable scale}	Total of sustainability rating (SR) – priority	SR of priority
									1	2	3	4	5		
	28		Quality of conferences, seminars and workshops: Number of key-notes speaker	√				1: 1–2, 2: 3–4, 3: 5–6, 4: 7–8, 5: More than 9							
	29		Quality of conferences, seminars and workshops: Level/status: (% of participant Local: International)				√	1: 10:90, 2: 20:80, 3: 30:70, 4: 40:60, 5: 50:50				4			
	30		Quality of Conferences, seminars and workshops: Quality of report				√	1: Low, 2: Intermediate, 3: Medium, 4: Good, 5: Excellent			3				
	31		Number of recognition (referral points, awards)				√	1: participant, 2: 4–5th place, 3: 3rd place, 4: 2nd place, 5: champion/special award					5		

Table 12.4 Community engagement for sustainability assessment

Priority setting	NO	Indicator	Output indicator	NA	SD Env	SD Eco	SD Soc	Description	Sustainable scale 1	2	3	4	5	Total of Sustainability Rating (SR) – priority	SR of priority
Community engagement	32	Community involvment	Number of community development projects		√	√	√	1: 1, 2: 2, 3: 3, 4: 5, 5: more than 7					5	46	1.8
	33		Number of population involved in the community development programmes		√	√	√	1: 5%, 2: 10%, 3: 30%, 4: 60%, 5: 80% and above			3				
	34		Number of households involved in the community development programmes		√	√	√	1: 5%, 2: 10%, 3: 30%, 4: 60%, 5: 80% and above			3				
	35	Partnership	Number of networking from third parties collaborating in the programmes: Public sector			√	√	1: 1, 2: 2, 3: 3, 4: 5, 5: more than 4		2					
	36		Number of networking from third parties collaborating in the programmes: Private sector			√	√	1: 1, 2: 2, 3: 3, 4: 5, 5: more than 5			3				
	37		Number of networking from third parties collaborating in the programmes: NGOs			√	√	1: 1, 2:2, 3:3, 4:5, 5: More than 6			3				

continued

Table 12.4 Continued

Priority setting	NO	Indicator	Output indicator	SD				Description	Sustainable scale					Total of Sustainability Rating (SR) – priority	SR of priority
				NA	Env	Eco	Soc		1	2	3	4	5		
	38		Number of networking from third parties collaborating in the programmes: community based organizations (CBOs)			✓	✓	1: 1, 2: 2, 3: 3, 4: 5, 5: more than 7		2					
	39	Socio-economic implications	Improvement in socio-economic indicators: Improved income			✓	✓	1: Low, 2: Intermediate, 3: Medium, 4: Good, 5: Excellent		2					
	40		Improvement in socio-economic indicators: Better employment opportunities	✓				1: Low, 2: Intermediate, 3: Medium, 4: Good, 5: Excellent							
	41		Improvement in socio-economic indicators: Better safety net				✓	1: Low, 2: Intermediate, 3: Medium, 4: Good, 5: Excellent		2					
	42		Improvement in socio-economic indicators: Availability of diversified income	✓				1: Low, 2: Intermediate, 3: Medium, 4: Good, 5: Excellent							
	43		Improvement in socio-economic indicators: Diversified markets			✓		1: Low, 2: Intermediate, 3: Medium, 4: Good, 5: Excellent		2					

44	Socio-cultural implications	Improvement in socio-cultural indicators (enhances socio-cultural values): Language and traditional customs	√	1: Low, 2: Intermediate, 3: Medium, 4: Good, 5: Excellent	2
46		Improvement in socio-cultural indicators (enhances socio-cultural values): Improved self-esteem and identity	√	1: Low, 2: Intermediate, 3: Medium, 4: Good, 5: Excellent	2
47		Improvement in socio-cultural indicators (enhances socio-cultural values): Improvement in health condition	√	1: Low, 2: Intermediate, 3: Medium, 4: Good, 5: Excellent	2
48		Improvement in socio-cultural indicators (enhances socio-cultural values): Reduce social problems	√	1: Low, 2: Intermediate, 3: Medium, 4: Good, 5: Excellent	2
49	Environmental implications	Improvement in environmental indicators: Increase in biodiversity	√	1: Low, 2: Intermediate, 3: Medium, 4: Good, 5: Excellent	1
50		Improvement in environmental indicators: Sensitivity for the environment	√	1: Low, 2: Intermediate, 3: Medium, 4: Good, 5: Excellent	2
51		Improvement in environmental indicators: Pleasant local environment	√	1: Low, 2: Intermediate, 3: Medium, 4: Good, 5: Excellent	1

continued

Table 12.4 Continued

Priority setting	NO	Indicator	Output indicator	NA	SD Env	SD Eco	SD Soc	Description	Sustainable scale 1	2	3	4	5	Total of Sustainability Rating (SR) – priority	SR of priority
	52		Improvement in environmental indicators: Increase environmental awareness		√	√	√	1: Low, 2: Intermediate, 3: Medium, 4: Good, 5: Excellent			3				
	53		Improvement in environmental indicators: Increase green practice		√			1: Low, 2: Intermediate, 3: Medium, 4: Good, 5: Excellent	1						
	54	Community achievement	Number of local champions that drive the community development from within				√	1: 1, 2: 2, 3: 3, 4: 5, 5: more than 7		2					
	55		Continuous link between AIU and the community even after the community has graduated	√				1: Low, 2: Intermediate, 3: Medium, 4: Good, 5: Excellent							
	56		Number of community that can be showcased to other new communities	√				1: 1, 2: 2, 3: 3, 4: 5, 5: more than 7							
	57		Number of community that can become consultants to other new communities	√				1: 1, 2: 2, 3: 3, 4: 5, 5: more than 7							

completed during the preparation of community immersion. Please refer to Table 12.3 for detailed calculation.

Table 12.4 presents community indicators with six main areas covered – community involvement, partnership, socio-economic implications, socio-cultural implications, environmental implications and community achievement. For community involvement, the output indicators are assessed based on numbers and monetary values, perceptions and impacts. The value of community indicators is 46, which, when divided by the total number of output indicators (26), gives the result of 1.8, which on the SRAS scale indicates low sustainability (see Figure 12.4). The community component is low because the project is still ongoing. Please refer to Table 12.4 for more detailed calculation. The overall, aggregated result shows that the sustainability rating for the AIU/SLA programme is 2.1, which is a medium level of sustainability (see Table 12.6).

Table 12.5 shows for the Kensiu project that the higher impact of the SD pillars is on the social dimension (67.66%), with both the environmental and economic dimensions at the same percentage (38.60%). Therefore, the percentage of ESD-SD is 48.29 per cent which is considered as a medium level of performance on ESD-SD.

Figure 12.4 shows the target based on the current result, whereas by Year 2, performance on the system is targeted at 3.3 and by Year 3 at 4.3. It is in this way that the system allows intervention and improvement to achieve

Table 12.5 SRAS result for Kensiu Project, Year 1 (%) in relation to ESD-SD

ESD-SD	Environment	%	Economic	%	Social	%
Teaching	6	46.15	5	38.46	8	61.54
Research	7	38.89	7	38.89	13	72.22
Community engagement	8	30.77	10	38.46	18	69.23
Total SD	21	38.60	22	38.60	39	67.66
% of ESD-SD				48.29		

Note: Low ESD-SD (1–30%), Medium ESD-SD (31–69%), High ESD-SD (70–100%)

Table 12.6 Sustainability rating for Kensiu Project, Year 1 (Scale)

Sustainability	Sustainable scale	Sustainability rating
Teaching	1.9	Low sustainability
Research	2.5	Medium sustainability
Community engagement	1.8	Low sustainability
Sustainability rating (overall)	2.1	Medium sustainability

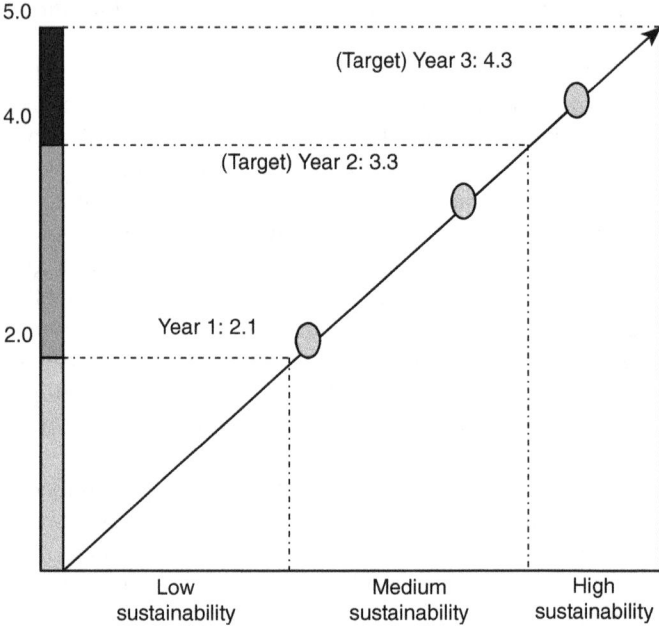

Figure 12.4 Sustainability rating for students' performance in the Kensiu project, Year 1 with targets for Year 2 and 3

sustainability goals. Individual group performances are manifested in the marks accumulated during both theoretical and practical assessments including in examinations and group assignments. The results are then translated as their cumulative grade point averages (CGPA) to reflect individual students' performance. In addition students are assessed in terms of their humane behavioural competency through the Humaniversity Competency Framework (HCF) using selected sessions mainly conducted during practical sessions. The group performances are manifested in the quality and quantity of students graduated with transferable skills, and are reflected in the number and quality of recognition indicators, such as awards, and the results of competition based on the quality of the practical projects. The course performance is determined by identifying the number of transferable skills that are applied by the students after graduation, including their self-initiative in community projects and employability after graduation. The performance of the SLA course is reflected in the whole university performance through the key performance indicators of teaching, research and community engagement. Teaching is conducted according to the ESD approach using formal, informal and non-formal methods such as media, interactive simulation and experiential learning on sites. For teaching, this includes recognition of the course by receiving accreditation from the Malaysian Quality Accreditation (MQA) body. Evidence of visibility and recognition of the teaching methods through media like television and internet and the consequent emergence of the programme as a referral point for community–university engagement is also taken into account. For research, the contribution is in its effort to promote transdisciplinary research, the number and quality of publications, grants, consultations, seminars, workshops and conferences related to SLA projects and activities. Indicators for community engagement involve assessing the quality, quantity and impact of the community projects. Evidence can include continuous rapport and the quality of networking with stakeholders and the support of the third parties (NGOs, public and private sectors) involved in the projects. The indicators are reflected in the quantity, quality and impacts of the community projects.

The strength of SRAS is that it is designed to monitor and evaluate the progress and performance of the SLA programme in the unique context of both AIU and Malaysia in ways that take into account the university's operating context and meet the mission and vision of the university to graduate academically qualified students and create disciplined, caring and giving individuals with a high level of humanitarian values. SRAS provides indicators and signals to meet goals through intervention and the assessment tools can be quickly sharpened and enhanced over time due to its 'bottom-up' nature. The 5-point Lickert scale is relatively sensitive. Finally, SRAS is designed not only to indicate existence of SD activities but also to evaluate impact and assessment which is done annually.

However, unlike the global SD ratings, SRAS is not designed for visibility, but for 'in-house' purposes only, hence there is no avenue to share ideas and successful SD strategies with other HEIs. SRAS also uses only quantitative values in the quality assessment process. SRAS covers both process and impact indicators. Process indicators are sometimes known as input indicators, that is, the process done to achieve impact indicators or outcomes. Both indicators need to be measured, as relying on impact alone is not suitable since impact indicators are only achieved at the end of the projects. Process indicators also help to facilitate the monitoring and evaluation process. Most of the indicators under teaching and research are process indicators, whereas for community engagement, the assessments are mainly focused on impact indicators, except for community involvement and networking categories. A final, substantial, weakness of SRAS is that it only assesses the SLA programme and not the SD activities of the university.

Conclusion

The sustainable assessment rating systems that have been developed worldwide are important to promote the continuous quality improvement of sustainable development practices in higher education institutions. The rating systems are similar in that most of them take into account the three pillars of sustainable development – socio-cultural, socio-economic and environmental dimensions – in their four targeted areas. The targeted areas are teaching; research; community engagement; and operations (which can also include governance). The case study of AIU shows that the key elements of the global rating systems can be applied at a local level with, in the AIU case, a particular focus on community engagement. To be recognized and gain global visibility in all these efforts, it is crucial for AIU to be assessed fairly against its own cohort, mission, context and identity: hence the relevance of a specialized ranking system focusing on sustainability and community–university engagement. For the SLA programme and the SRAS rating system to realize their full potential, continuity of the philosophy, vision and mission of the university should be ensured. Foundation values and an understanding of the importance of promoting ESD is essential at all levels, from funders and leaders to those involved in implementation. Hence the need for strong political will, leadership, entrepreneurship, networking and the quest to learn new skills, adopt new paradigms and break away from a silo research mindset to ensure the successful implementation of both the ESD programme and the rating system.

References

Chambers R., Conway G. R. (1991) *Sustainable Rural Livelihoods: Practical Concepts for the 21st Century* (Brighton: Institute of Development Studies, University of Sussex).

Economic Planning Unit, Malaysia (2006) *9th Malaysian Plan (2006–2010)* (Malaysia: Office of the Prime Minister).

EUA Report on Rankings (2011) 'Global University Rankings and Their Impact', http://www.eua.be/pubs/Global_University_Rankings_and_Their_Impact.pdf, date accessed 27 April 2013.

Senaha E. (2010) 'Alternative University Appraisal Based on ESD. Perspective on University Performance Evaluation on March 16, 2010', ias.unu.edu/resource_centre/Eijun%20Senaha.pdf, date accessed 25 June 2013.

Shriberg M. P. (2002a) 'Sustainability in US Higher Education: Organizational Factors Influencing Campus Environmental Performance and Leadership', http://promiseofplace.org/research_attachments/Shriberg2002SustainabilityinHigherEdu.pdf, date accessed 13 June 2013.

Shriberg M. P. (2002b) 'Institutional Assessment Tools for Sustainability in Higher Education: Strength, Weakness and Implication for Practice and Theory'. *International Journal of Sustainability in Higher Education*, 3 (3) 202, 254–270.

Sustainable Endowments Institute (2011) 'The College Sustainability Report Card', http://www.greenreportcard.org/report-card-2011/executive-summary, date accessed 27 April 2013.

ULSF (Association of University Leaders for a Sustainable Future) (2012) 'Sustainability Assessment Questionnaire', http://www.ulsf.org/programs_saq.html, date accessed 27 April 2014.

UNESCO (2010) 'Rankings and Accountability in Higher Education- Unesco', http://www.unesco.org/new/fileadmin/MULTIMEDIA/HQ/ED/pdf/RANKINGS/ Downing.pdf, date accessed 25 June 2013.

UNESCO Forum on Rankings and Accountability. 'What's the Use of Ranking? Using Rankings to Drive Internal Quality Improvements', www.unesco.org/new/fileadmin/.../HQ/ED/.../RANKINGS/Downing.pdf, date accessed 27 April 2013.

UNESCO World Conference on Education for Sustainable Development (2009) 'Bonn Declaration', http:// unesdoc.unesco.org/images/0018/001887/188799e.pdf, date accessed 27 April 2013.

Index

accountability, 2, 7, 27, 34–35, 40, 117, 119, 120, 122, 126, 132
approach
bottom-up, top-down approach, 3, 51, 88, 90, 105–106, 126, 245
holistic approach, 10, 50, 108, 122, 127, 149
participatory approach, 137, 210, 211
problem- and project-based learning (PPBL) approach, 197, 199, 201, 204, 222
quality culture approach, 55–56, 59, 61
sustainable livelihood approach, 5, 261–266
transdisciplinary approach, 73, 146, 149
whole (of) institution approach, 8, 10, 11, 14, 19, 50, 62, 70, 74, 115, 121, 125
autonomy (of HEI), 16, 25, 89–91, 114, 119, 120, 122, 126, 131, 133

Bologna process, 17, 91–100, 105, 107–110, 115–116, 119

competence(s)/ capability(ies), 7–10, 14–15, 61, 66, 72, 77, 99, 118–119, 141, 148, 149, 151, 186, 198–201, 203–213, 216–217, 219–220, 223, 224, 237, 239–242, 263

indicator(s), *see also* performance
qualitative and quantitative indicators, 15, 123, 132

knowledge
knowledge production, 33–34, 41
knowledge production accountability, 35
knowledge role, 6–7, 8, 25, 32

leadership, 11, 13, 14, 49, 54, 66, 67–75, 78–79, 81, 117, 154, 163, 164, 230, 242, 249–250

management, *see also* quality
change management, 2, 3, 4, 5,67, 87
strategic management, 18, 20, 121, 122–123, 125, 138

participation, 8, 14, 51, 55–57, 61, 67, 69, 75, 80, 135, 136, 139, 146, 155, 159, 165, 168, 171, 180, 190, 267
performance
performance agreement(s), 17, 80, 90, 92, 93, 95, 97–99, 103–107, 110, 131, 133, 138–139, 146
performance indicators, 38, 77, 125, 234, 263, 266, 278

quality, 1, *see also* quality assurance
quality assessment, 2, 16, 17–19, 67, 75, 78, 80–81, 99, 120, 122, 132, 140, 263, 279
quality criteria, 68, 77, 79, 99, 118, 122–123, 128, 132
quality culture, 12, 18, 51, 53–59, 61–62, 119, 132, 135
quality management, 2, 3, 15, 17, 19, 20, 51, 54, 57, 62, 67, 75–77, 81, 89, 91–92, 94, 99, 106–109, 123, 125, 131–141, 146, 148–149, 230, 235
quality assurance, 2, 3, 16, 18–20, 52–53, 99, 108, 119–122, 149, 203, 215
critical factors, 245–247
external/ internal (system), 16–17, 119–120
institutional/ internal, 122–125, 127

ranking(s), 13, 15, 25, 26–30, 36–38, 40, 42–44, 75, 164, 256, 280
reporting, 78, 89–90, 138, 155–156, 158, 163, 165, 168, 258

sustainable development/ sustainability, 115, 116, 126
  Education for sustainable development, 49, 56–58, 72, 91–92, 100–104, 106, 109, 141–142, 185–187, 198–203, 210–211, 213, 215, 219–220, 223, 244–245
  Sustainable development in higher education, 8, 41, 42, 50, 67–70, 71, 100–106, 141, 153, 192–193, 255
  (Sustainability) science/research, 7, 29–31, 33–35, 57, 74, 103–104, 149, 192–193, 201–202, 223

Transformation towards sustainable development
  by higher education institution, 1, 3, 50, 70–75, 79–80, 103–105, 115, 121, 159, 160, 171, 193, 197, 201, 232
  principles, 69, 81, 125–127
  role of education, 10, 100–101
  role of quality assurance, 3–4, 15–16, 19–20, 67, 75–79, 81, 94–99, 107

values, 10, 13, 18, 53, 55–56
  critical factors, 245–247
  external / internal (system), 16–17, 119–120
  framework, 235–238
  institutional/ internal, 122–125, 127
  as learning, 149, 203, 215
  as transformation,18
  for transformation, 18–20

GPSR Compliance
The European Union's (EU) General Product Safety Regulation (GPSR) is a set of rules that requires consumer products to be safe and our obligations to ensure this.

If you have any concerns about our products, you can contact us on

ProductSafety@springernature.com

In case Publisher is established outside the EU, the EU authorized representative is:

Springer Nature Customer Service Center GmbH
Europaplatz 3
69115 Heidelberg, Germany

www.ingramcontent.com/pod-product-compliance
Lightning Source LLC
Chambersburg PA
CBHW071615100426
42873CB00004B/54